EFCE Event No. 306

# Ergonomics Problems in Process Operations

*Organised by the Institution of Chemical Engineers in association with the Ergonomics Society. Held at the University of Aston in Birmingham 11-13 July 1984.*

**Organising Committee**

| | |
|---|---|
| D. Whitfield *(Chairman)* | Ergonomics Development Unit, University of Aston in Birmingham |
| P.K. Andow | Department of Chemical Engineering, University of Technology, Loughborough |
| L. Bainbridge | Department of Psychology, University College, London |
| K. Duncan | Department of Applied Psychology, U.W.I.S.T., Cardiff |
| B. Eddershaw | Petrochemicals Division Headquarters, ICI plc, Wilton, Middlesbrough |
| D.E. Embrey | Consultant, 1 School House, Higher Lane, Dalton |
| A. Faulkner | Organics Division, ICI plc, Huddersfield |
| A. Hale | Department of Occupational Health & Safety, University of Aston in Birmingham |
| D. Lihou | Department of Chemical Engineering, University of Aston in Birmingham |
| I.G. Umbers | Warren Spring Laboratory, Stevenage |
| S. Wood | Digital Technology Centre, The Lummus Co Ltd, Northampton |

**INSTITUTION OF CHEMICAL ENGINEERS**
**SYMPOSIUM SERIES NO. 90**

**ISBN 0 85295 172 8**

**PUBLISHED BY THE INSTITUTION OF CHEMICAL ENGINEERS**



First edition 1984 – ISBN 0 85295 172 8

MEMBERS OF THE INSTITUTION OF CHEMICAL ENGINEERS (Worldwide) SHOULD ORDER DIRECT FROM THE INSTITUTION

Geo. E. Davis Building, 165-171 Railway Terrace, Rugby, Warks CV21 3HQ.

*Australian orders to:*

R.M. Wood, School of Chemical Engineering and Industrial Chemistry, University of New South Wales, PO Box 1, Kensington, NSW, Australia 2033.

*Distributed through the world (excluding Australia) by Pergamon Press Ltd. except to IChemE members.*

| | |
|---|---|
| U.K. | Pergamon Press Ltd., Headington Hill Hall, Oxford OX3 0BW, England |
| U.S.A. | Pergamon Press Inc., Maxwell House, Fairview Park, Elmsford, New York 10523., U.S.A. |
| CANADA | Pergamon Press Canada Ltd., Suite 104, 150 Consumers Rd., Willowdale, Ontario M2J 1P9, Canada |
| FRANCE | Pergamon Press SARL, 24 ru des Ecoles, 75240 Paris, Cedex 05, France |
| FEDERAL REPUBLIC OF GERMANY | Pergamon Press GmbH, 6242 Kronberg- Taunus, Hammerweg 6, Federal Republic of Germany. |

*British Library Cataloguing in Publication Data*

Ergonomics problems in process operations. – (EFCE event; no. 306) – (Institution of Chemical Engineers Symposium series, ISSN 0307-0492; no. 90)

1. Process control    2. Human engineering
I. Institution of Chemical Engineers
II. Ergonomics Society    III. Series    IV. Series 670.42'7    TS156.8

Pergamon Press ISBN 0-08-030282-3

Library of Congress No: 84-16686

# PREFACE



The engineering world cannot afford to ignore ergonomics – no longer limited to academic research but directly applicable to process operations and design. As the pace of technology steps up, especially in computer applications, so attention to the human factor becomes imperative.

The objectives of the symposium are: To enable engineers from process, power, and offshore industries to exchange information and experience with ergonomics specialists; To discuss ergonomics problems in process operations and methods of recognising and controlling these problems; To explore effective ways of applying ergonomics research to solving process operations problems.

The symposium is intended to appeal to: Operations and maintenance managers: Equipment designers and project engineers: Designers and managers of pilot scale R & D plants: Ergonomists.

# CONTENTS

**3. Job Design, Organisation and Training**

# ELEMENTS OF PROCESS CONTROL OPERATOR'S REASONING : ACTIVITY PLANNING AND SYSTEM AND PROCESS RESPONSE-TIMES.

Maud Boël (*), François Daniellou (**)

Abstract

Process industries are particularly characterized by dynamic aspects of process evolution.

The operator's activity in the control room of a computer-controlled process has to take into account different time-characteristics : some of these are specific to an automatized process (response-time of analyzers, "age" of informations); others are not (such as process delays, response-time of field informations).

Some characteristics of the operator's reasoning, issued from research carried out over one year in the control room of an oil-refinery are presented :
- on the one hand, how the operator plans his activity when facing one incident, and manages the different response times,
- on the other and, how the operator faces several simultaneous incidents.

The elements presented in this paper issue from research carried out in the control room of a refinery, the process of which had been computerized a few years ago.

The control room operators control the process on the basis of :

. information displayed on visual display units (V.D.U.)
. information collected from the plant by field operators.

Based on the example of one incident, some characteristics of the operator's reasonings when he is facing incidents will be presented, with special reference to how he takes into account the different response-times induced by the system, and the way he deals with simultaneous incidents.

## I - THE METHODS USED

It is obviously very difficult to analyse the operators' mental process in a real control-room situation as there is no question of interrupting them in

(*)Laboratoire Communication et Travail, Université de Paris-Nord, Avenue J.B. Clément - F 93430 VILLETANEUSE.

(**) Laboratoire de Physiologie du Travail et d'Ergonomie du CONSERVATOIRE NATIONAL DES ARTS ET METIERS, 41, rue Gay-Lussac,F75005 PARIS.

the course of the resolution of an incident.

. The method chosen was to continuously record (*) the control operators' behaviour, that is :

- the consulted displays and the actions taken,
- the professional conversations and the radio communications and to submit some of these records later to the operators during interviews in which they were asked to explain the reasons for the different actions carried out from the board.

The confrontation of :

- the actions recorded,
- the verbalizations made during the activity (radio communications, conversations in the control room or comments to the researchers)
- operators' subsequent comments

allows restoration within some limits, of elements of the reasonings supporting the action at one moment.

It is obvious that the methods used do not make it possible to infer the structure of the operator's long-term knowledge of the process control.

The operators concerned are 39 to 47 years old. They have been in the post from 2 to 5 years.
All of them had been field operators for some years in the past.
When appointed to the control-room, they were trained in dual post for four to eight months.

## 2 - AN EXAMPLE OF AN INCIDENT

To introduce some elements of the operator's thought processes when facing an incident, an example will first be presented : owing to the number and diversity of indications consulted, emphasis is not laid on a quantitative approach, but rather on the organization of the observed VDU and field consultations.

The word "incident" refers to any disturbance or adjustement requiring the team's intervention.

To understand the example, it is necessary to know that the operator has to type the number of the desired displays to set the items of information he is looking for. The appearance of a new display erases the previous one.

. The example :

Two refining columns, C1 and C2 are linked together in two ways :

. on the one hand, the top products of C1 feed C2;
. on the other hand, a reflux of C1 is used to heat C2.

---

(*) The records were made during a total period of 105 hours, in 39 different shifts, by means of a V.T.R., tape-recorder or note taking.

During one of the recorded incidents, the C1 reflux circulating pump broke down. Such an incident can lead to :

- a general disturbance in C1 as it is no longer cooled enough, which means that out-of-tolerance products are sent to C2;

- a lack of heating in C2, and the risk of feeding downstream units with faulty products.

It has been observed that, in the first moment, the operator seeks to restore normal temperatures in C1. He acts on from one to eight regulation-loops relating to C1, while only keeping watch on the temperature in C2. From the evolution of this indicator and from field information about the permutation of the pump, he is led to switch over to a new set of groups (i.e., displays of the automatic regulators) which is made up of the connections between C1 AND C2. He then acts on the loops which regulate these connections, only keeping an eye on the developments in C1 through a small number of indicators. At the same time, he watches for possible consequences dowstream through only one source of information, the level of the C2 top tank. If this level changes, this means that the disturbance is about to reach downstream units, and the operator will then consider a new set of groups which includes C2 and the next unit.

In this example,the following can be observed :

. the operator works on sets of groups, synopses and graphs, relating to :

. first, column C1,
. then, columns C1 & C2,
. then these columns plus another unit,

. the sets of groups change in the course of the incident.

. While working on one set, the operator consults certain other indicators (e.g., the temperature of C2 while working on C1).

. switching from one set of values to another seems to be the result either of a change in the disturbance (change in the tank-level) or of the arrival of information from the field, which involves delays (waiting for a pump to be exchanged requires 3 minutes while other actions only stretch over 48 seconds).

. Through the verbalizations made, it is learned that the operator determines a temporary objective : not to lose the reflux flow.

The incident presented above is only one illustration of the activity observed during incidents.

Despite the variability of incidents and of operators controlling them, an interpretative model presented below seems to be applicable to the analysis of most incidents observed, as far as the overall organization of information processing is concerned.

## 3. ELEMENTS OF A MODEL

At one time, the operator has to deal at one time with more or less numerous incidents which progress simultaneously. These incidents may affect different parts of the refinery, and possibly interfere with each other.

To solve an incident, the operator determines intermediate objectives to be achieved. Theses objectives are fixed on the basis of a representation of the incident issued both from immediately available information and from knowledge gained through experience.

To reach such an objective, the operator focuses on specific parts of the refinery, which are either directly affected by the incident or functionally linked to the sections concerned. The action taken to reach an objective concerns what will be called a Set of Functionally Linked Elements (S F L E), defined in detail below.

The following paragraphs deal with :

. Some features of the operator's reasoning within one given SFLE,
. his changing from one SFLE to another in the course of one incident,
. **his facing several incidents at** one time.

### 3.1. Reasoning Within One SFLE

Owing to the process design, disturbances in one place are likely to have many consequences in several different areas of the refinery. The parameters that may be affected are thus extremely numerous.

Consequently, the question to be investigated is : how does the operator organize the way he keeps watch on this number of parameters ?

It can be noticed that he partly foresees the likely developments of the incident, based on his knowledge of the process and of similar incidents which had occurred in the past. He is more or less in a position to anticipate :

- the nature of oncoming disturbances,
- their occurrence order.

He can then determine intermediate objectives, which aim to limit the foreseeable sequence of repercussions of the initial disturbance.

Thus partitioning the resolution of the incident, he can focus his current action on a subset much smaller than the set of all parameters likely to be affected. This subset, which will be referred to here as a "Set of Functionally Linked Elements", includes the details of all parameters involved in reaching the intermediate objective. It is the framework within which the operator keeps watch on parameters, acts on the process (especially valves), either straight from the keyboard or through field operators.

It has to be noted that this subset is functional, meaning that it can include parameters relating to different refining columns or other devices linked by the process.

Since the operator's reasoning is orientated by the consequences of the incident (at least the ones he knows), the SFLE is constructed on the basis of the units likely to be reached by the incident.

Therefore, the SFLE is not mainly specific to the causes of the incident, but rather to the foreseeable consequences which the incident may have on the other parts of the refinery.

### 3.2. Switching From One SFLE To Another.

As indicated above, the development of an incident is only partly foreseeable. The operator, while focusing on an intermediate objective, may be led to modify his plans by the occurrence of a new event or by the abnormal development of a parameter. He then determines a new intermediate objective, which causes him to work within the frame of a new SFLE. The question is then to understand how, while working within SFLE 1, he decides to leave SFLE 1 and devote himself to a new SFLE, say SFLE 2.

As already mentioned, the operator can anticipate that the incident is likely to affect not only SFLE 1, but also other parts of the plant. It can be noted that, while closely following the development of the situation in SFLE 1, he keeps watch on the evolution of these other parts. But the monitoring maintained on these parts is not of the same nature as that for SFLE 1.

Whereas within the current SFLE the operator follows all parameters in a nearly continuous manner, he keeps an eye on the development of other parts only through the intermediary of a restricted number of parameters, that will be referred to as "CUES". These cues give information in a synthetic way on the overall state of a unit. If one of them becomes abnormal, the operator leaves SFLE 1 to focus on the detail of the new problem arising, within a new SFLE, SFLE 2. He will in turn keep an eye on the next developments in SFLE 1 through a limited number of cues.

. Cues and memorization

It seems that watching neighbouring units through a limited number of cues permits the operator to reduce the number of continuously followed parameters, which may perhaps lead to a reduced memorization.

But on the other hand, cues are not sufficient for precisely identifying components of a malfunction and, when necessary, they have to give way to the detailed examination of the relevant SFLE.

### 3.3. Several Incidents

In the reality of work, the operator often has to face several incidents developing at the same time.

It can be observed that, as above, the operator will focus on one SFLE relating to one incident and keep an eye on its other components AND on other incidents through cues.

Managing incidents requires an up-dating of the representations at every occurrence of a new event. It implies memorizing developing incidents while dealing with one of them.

### 3.4. Responses Times

It has been indicated before that switching from one SFLE to another was a result of a parameter developing abnormally. Another factor has to be taken into account, namely the operator's knowledge of the response-times of the process and the system.

The main features of response times are described below and their consequences for the operator's action discussed.

. Response-time of field interventions

As described in another paper (see : DANIELLOU, F., BOEL, M. in this Symposium) all items of information required by control room operators are not available from VDUS. Some have to be collected in the field by field operators. Other indications have to be confirmed by field checking. In the same way, all actions on the units cannot be taken from the keyboard. Some require an intervention on the spot, e.g., starting or stopping a pump.

In all cases, asking for information or intervention from the field implies a certain response-time, due to the operator's displacements or actions. Thus some devices fans are situated on top of a refining columns, some valves are difficult to operate and so on. The time required for operating a device is thus specifically dependent on its implantation.

It can be observed that control room operators are able to estimate the time required by the outer operators to satisfy their demands. This is probably possible mainly because they have learnt to know the units when they were operators in the field.

. Response-times of analysers

Automatic analysers display the composition of products. However, according to the type of analysis, the response time range from a few seconds to ten minutes or more.

In some cases, the operator who changes an adjustment will know its result only after waiting for the measure to be completed.

. Process delays

The flows of products between the units take time, sometimes an hour or more. The operator knows this and he can evaluate how long a disturbance will take to reach the next unit.

. Consequences of response-times on operators' thought process

The knowledge of these different kinds of response times is a major element of the operator's management of one or several incidents. Once he has given rise to an action, either from the keyboard or through a field operator, he is often observed to provisionally leave the SFLE with which he had been busy, for as long as the response-time mentioned is not over. From the operators' verbalization, it has been learned that he switches back to it when he has the feeling that the requested actions have had time to take effect. In such a case, the change in SFLE is not necessarily brought about by the evolution of a cue, but rather by the awareness of elapsed time.

It seems that while processing in the mean time information related to other problems, the operator maintains and additional activity of time evaluation.

It can be noted that this evaluation is extremely accurate when there are not too many mean-time incidents.

But on the other hand, the evaluation can be troubled when incidents occur which require long or complex interventions.

3.5. Summary :

The operator's activity is organized within sets of functionally linked elements, determined in relation to intermediate objectives that have to be reached to solve an incident. While acting within one SFLE, the operator keeps watch on cues relating either to other aspects of the incident, or to other incidents, and he evaluates the time required by the actions taken. He switches from one SFLE to another

. because a new event has occurred,
. because of the evolution of a "cue",
. or because a predetermined length of time has elapsed.

This description of the operator's activity makes it possible to put forward a few indications on information display.

## 4. CONCLUSION

An analogy can be evoked between the organization of information processing as described above, and the well known organization of visual function :

. on the one hand, a kind of a "central vision", the current SFLE, where attention is paid to the detailed state of parameters,

. on the other hand, a kind of a "peripheral vision" where synthetic indicators are watched; a sudden change, or a threshold being crossed by one of these "cues", causes a "movement" of focusing on a new central SFLE.

It is quite obvious that this analogy is merely descriptive, without any physiological substratum at all having been included.

The results are limited, and many points have to be clarified, such as how SFLES are organized from experience, the mechanisms of time evaluation, how priorities are determined and so on.

It seems however that, at the present time, it is possible to highlight the consequences of this description of activity in terms of information display design, which has to take into account all of following items :

- working within SFLE : this is obviously the role of synoptic displays, which cannot be designed without posing questions about how parameters are combined when solving one incident.

- monitoring "cues" : - these must be available without any complex search for information. They cannot merely be elements to be extracted from among all other parameters in a synoptic display.

. evaluation elapsing time : this can be aided by suitable devices.

It is obviously essential to validate, give greater precision to and enrich this model in other automatized processes, especially in the chemical industry. Research currently being carried out should permit some progress.

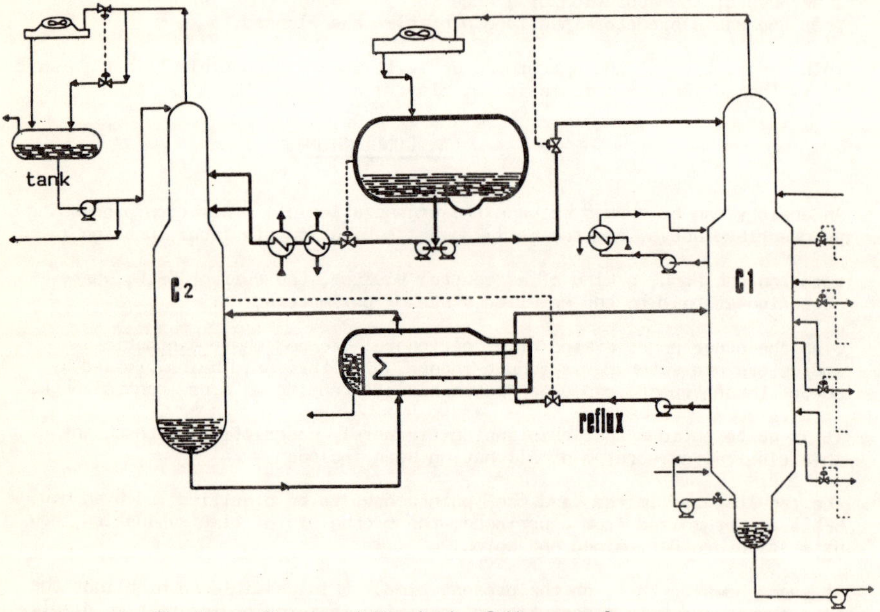


The two columns and the tank of the example.

# ERGONOMICS OF THE CONTROL OF SIMULTANEOUS PROCESSES: CASE STUDY IN BIOCHEMICAL INDUSTRY

J. LEPLAT * & M. ROCHER **

The simultaneous control of several processes running in a non-synchronized manner requires a time-sharing-based activity of this type has been investigated in a biochemical manufacturing plant. After a brief presentation of the task some essential aspects of the activity have been highlighted. A still summary preliminary analysis revealed the important role of the anticipation and highlighted some of these conditions. Ergonomic consequences of this work have been investigated which should result especially in reducing the temporal demands and the incidents.

## INTRODUCTION

In some even technically advanced industrial systems, the operators are in charge of the control of processes running simultaneously in a non-coordinated manner. Their activity is thus based on "time sharing" since it has to be distributed between the different processes depending upon the requirements of the moment. This characteristic which constitutes currently the main difficulty will be investigated here by means of a study carried out in a biochemical manufacturing plant. This study is extracted from a wider work performed by one of us (Rocher, 1982) (6) and dealing with the process control in biochemical manufacturing facilities. Its specific goal was not the subject which will be dealt with here. This subject however appeared as being of major importance in the analysis of the operator activity and work schedule. These circumstances explain also the character of the present work. It is not a systematically organized research on this subject, likely to bring a rigorous support for some hypotheses: it is rather a whole of observations capable of revealing the importance of the raised problem and likely to justify and to direct further research work on the activity dealing with the control of simultaneously running processes.

The situation which will be investigated here belongs to a very wide category embracing various types of work: supervision of several apparatuses; switchboard operation at an emergency station; nursing (Theureau, 1981) (8); executive tasks, etc. In all these cases multiple tasks are to be performed at the same moment, so that an order of execution must be set -this order itself being likely to be challenged if another urgent task occurs unexpectedly.

* Directeur du Laboratoire de Psychologie du Travail - EPHE - 75005 PARIS.
** Chargé d'Etudes - ROUSSEL UCLAF - 93230 ROMAINVILLE - FRANCE.

The analysis of the planning and coordination of the individual as well as group activities in this type of work is of much theoretical importance. It leads to question about the mechanisms governing the planning of the activity. For the ergonomist, the knowledge of such mechanisms should lead to a technical and organizational adjustment of the work. With Savoyant (Leplat and Savoyant, 1982) (3) one of us proposed the bases of a reference framework within the limits of which the said activities could be investigated, laying emphasis on the characteristics to be considered in the planning activities. It may be noted some relationship between these problems and those treated in the 1950's by Conrad (1954) (1) and Mackworth and Mackworth (1956) (4). They suggest interesting hypotheses on the mechanisms of the planning, namely by showing the widening of time field and the role of the anticipation which leads not only to respond to a brief occurrence but also to create appropriate conditions for the easy treatment of the next occurences.

## I - Technical and Organizational Conditions

The work conditions which deal with the simultaneous operating processes are presented in this first part and are to be related in the next part to the corresponding activity characteristics.

1- Presentation of the investigated workshop. The study has been carried out in the Fermentation Department of a biochemical manufacturing plant.

a) Description of the fermenter: the fermenter is a piece of equipment operating under steam pressure -thus under sterile conditions- and allowing the culture of the bacteria synthesizing the desired product. This apparatus is fitted with a central stirrer and a cooling system.

In each fermenter a continuous process is run on the basis of an about one-week cycle. The fermenters and the corresponding processes are independent of one another.

In the investigated workshop (34m x 12 m), the operator conducts 12 fermentors simultaneously, i.e., 12 processes which are not synchronized. To each several-meter-high fermenter correspond three working levels: ground, 1st and 2nd floor, the latter being the most used.

b) Process control room: this production system is partly automated. The automation has been introduced progressively within several years. It is partial because it applies only to some process phases for which it is not adapted to treat too great deviations from the expected operation.

The specific data relating to each fermenter are gathered in the control room, in this way two sources of information -theoretically identical- are available for the operator in both workshop and control room. But for "manual" phases, the intervention (e.g. valve operation) can be made only on the fermenter itself.

2- General description of a fermentor process.

a) Phases of the process: they are summarized in Table I. Within a given process the phases have different durations. In addition, the same phase has a duration varying from one batch to another.

TABLE I – Phases of the process – Main sources of organizational (O) and technical (T) variations of the phase durations of process.

| | PHASES | Duration range of phases and time break (Mean Duration) | Regulation Type | Operator task | Source of organizational (O) and technical (T) variation | |
|---|---|---|---|---|---|---|
| 1 | Reception of nutrient medium | 0 h 15 to 3 h 30 (0 h 40) | M (Manual) | To receive the nutrient medium from the service in charge of preparation. | O | Availability of the upwards service in charge of preparing the medium : time interval between the transfer of the 1 st part and the 2nd part of the medium. |
| 2 | Medium heating | 1 to 2 h (0 h 50) | M | To bring the temperature to the prescribed value (by progressive increase). | | cf. Incidents – Table II |
| 3 | Medium sterilisation | (constant - Not to be disclosed) | M | To keep the temperature at a constant value for an accurate time interval. | | Strict duration prescribed (Not to be disclosed). |
| 4 | Medium cooling | 1 h 30 to 4 h (4 h 35) | (Manual - Automated) | To bring the temperature to the prescribed value (by progressive decrease). | | cf. Incidents – Table II |
| 4 a | Time break | 0 to 11 h (2 h 55) | | | T<br>O | Time required by the preculture equipment to reach the prescribed quality specifications (decision to be made by the engineers).<br>Availability of the foremen (who can advance or detaly this phase by about 1 h). |
| 5 | Medium inoculation | 0 h 50 to 1 h 15 (1 h) | M | To transfer a preculture (from a little fermenter) to the fermenter containing the sterile nutrient medium) Operation to be performed with the foreman. | T and O | May vary depending upon the length of piping between the tank to be inoculated and that containing the inoculum. |
| 6 | Culture | 160 to 220 h (8 days) | A (Automated) | To check the parameters for differences from prescribed values. To prepare an additive solution for transfer to the fermenter. | T | Laboratory results and cf. Incidents Table II |
| 6 a | Time break | 0 h 30 to several hours (**) (a few hours) | | | O<br>O<br>O | - Presence of downwards service.<br>- Possibility for downwards service to store the fermentation broth (difference in hours of work).<br>- Availability of the operator. |
| 7 | Harvest of fermentation broth | 1 h 50 to 2 h 10 (1 h) | M | To transfer the fermentation broth to extraction department (downwards). | T | Size of the draining valve. |
| 7 a | Time break | 0 h 30 several hours (a few hours) | | | O<br>O | - Harvest hour of the previous process.<br>- Availability of the mechanician (in charge of removing the pH electrode). |
| 8 | Fermenter cleaning | 0 h 15 to 2 h (1 h) | M | To leave a clean fermenter for the next process. | T | Cleaning type depending itself upon the presence or the absence of a contimination from the previous culture or upon a major mechanical repair. |
| 8 a | Time break prioto next preparation | 0 to 14 h 30 (4 h 35) | | | O | Availability of upwards service (difference in work hours). |

The time intervals between the sequences may also largely vary from one process to the next (within the same 8-hour shift) and from an 8-hour shift to another. The above stated variability sources pertain on the one hand to technical and organizational demands (Table I) and, on the other hand to demands related to incidents (Table II).

TABLE II - Main Sources of Variations Related to Incidents (for 3 phases of the process).

| PHASE | Type of incident | Approximate frequency | Severity | Approximate time for recovery |
|---|---|---|---|---|
| Medium heating | -Insufficient steam flow-rate | 0 to 2 times per 8-hour shift | Medium | 5 min |
| | -Abnormal temperature | 2 to 3 times per 8-hour shift | Medium | 15 min |
| Medium cooling | -Insufficient cold water flow-rate | 0 to 2 times per 8-hour shift | Low | 5 to 15 min |
| | -Abnormal temperature | 2 to 3 times per 8-hour shift | Low | 15 min |
| | -Abnormal temperature of the discorded water | 0 to 3 times per 8-hour shift | Medium | 15 min |
| Culture | -Abnormal parameters, e.g., contimination of fermentation broth by external germs, pH, pressure and flow-rate foam level | 2 to 3 times per 8-hour shift | Medium or Low | 5 to 15 min |

b) Demands related to time: the previous demands (i.e. those related to the technique, to the organization or to the incidents) are reflected to the operator in terms of temporal demands. The development of the phases is required to be run at acceptable times and within acceptable time intervals, otherwise irreversible consequences could result, for the product. To respond to these demands the operator can make use of the allowed tolerances and time intervals. Among these restraints a distinction can be made between those resulting from the planning set up by the management and those resulting from hazards affecting this planning.

Time related demands affecting the official planning. The management tried to control the time-related demands by setting up regulations as well as a planning:
- Permanent regulations or directions. They remind us of the major time-related demands. These demands result mainly from the manufacturing procedure which prescribes in particular the sequential order for the development of the phases; the time at which these phases have to be initiated; their duration as well as the corresponding tolerances.
- Planning established on a one-week basis and indicating the starting hour of the phases of eacn process and the development of which is checked by the

end of each shift. Mainly those operations considered as priority tasks are mentioned in this planning.

Time related demands resulting from hazards. The planning is based on the smooth development of the process. Now the occurence of numerous unexpected hazards introduces some time-related variability in the development of the process. Let us mention a few major sources of hazards:
- Hazards related to the nature of the process. Let us remember of the biological nature of the process which is not totally controlled. Therefore, a high precision in the regulation of the parameters cannot be expected from the operator. The enormous volumes contained in the fermentors and the longlasting cycles result in producing phenomena of inertia which complicate the regulation operations and affect the prediction of their results.
- Hazards related to the organization. The achievement of some phases is related to the intervention of other services which are not always able to operate at the appropriate moment (e.g., because of differences in hours of work thus causing the execution to be delayed.
- Hazards related to the occurrence of incidents. The incidents affecting the development of a phase require more or less long-lasting and sometimes repeated interventions. In some cases, such incidents can result in stopping the process.

Means available for the operator to respond to the demands. As will be shown further, the operator is able to respond to the demands in a symptomatic or in a curative manner. Symptomatic manner: the major available means for the operator is to make use of the time-related tolerances: to increase or to reduce a time interval; to delay the initiation of a phase or the treatment of an incident. Curative manner: to act on the sources of hazards likely to cause unexpected demands the operator has only few available means. The biochemical process as well as the organization in its most part -especially the action on neighboring services- are beyond his control. Only remain available for him the means of anticipating the hazards, mainly the incidents, and then of taking in time some appropriate preventive measures.

## II - Analysis of the Activity

After having shown the characteristics of the investigated task we will examine how the operators meet the requirements of this task. To this end, we will try to determine some analytical ways likely to serve as a guideline in investigating operations of the same type.

1- Analysis of the operator movements. A direct consequence of the control of multiple pieces of equipment by the same operator is the importance of the movements required to go from one apparatus to another (between them 5-6 meters). The figure 1 proposes a representation method for these movements: it has been determined from observations made on the intervention sites of the operator by following him within the investigated time period. Any station near the fermenter aimed at checking a variable and/or at operating a control may be considered an intervention. The major part of the interventions take place on the same floor (2nd floor).

From these operations, it has been noted that the interventions during a 8-hour shift take place mainly on 2 or 3 fermenters, the others being less concerned by the activity. Moreover, the responsible persons try always to plan the production in such a manner that not more than 2 or 3 fermenters are concerned simultaneously with phases requiring many interventions.

The comparison between the diagrams of the same type as Figure 1 shows that there is no regular cycle and they reveal a high variation in the intervention sites and times within a shift and from a shift to another. Such a variation is a sign as well as a consequence of the impossibility for the operator to regulate his activity by repeating the same type of rounds. He must base this activity on the knowledge of the status of each process at a determined time.

2- Deviations between real and expected tasks. By examining the work processing it appears that the observance of the planning is only approximative. Figure 3 indicates the variations in the medium reception and inoculation hours. In this case for example, it seems that the reception of medium tends to be brought toward if the inoculation is likely to be made earlier. In fact, these inoculations are delayed very often. It appears that, when it is feasible, the operator brings the system in the best possible conditions to avoid any delay which could be attributable to his department. Such a phenomenon is particularly obvious with this critical operation (in the meaning of PERT method). It will be dealt with once more during the study of the anticipation. As a general rule, it appears that the expected times represent only fixed points around which the real work is organized.

3- Incident-related interruptions of activity. Figure 2 gives within a 3-hour period the intervention sequence corresponding to normal processing and to recovery from incidents. This figure shows that the incidents are frequent: the interventions they require represent about one-fourth of the total number of interventions. Such incidents are expensive on two accounts for the operator. First, they represent additional work and then, they disturb the normal processing of the activity; they may require from him a re-programmation of this activity. That is the reason why the operator will do his best to reduce the costs related to these incidents by trying to anticipate and to settle them preventively.

4- The anticipation and its symptoms. The analysis of the task has revealed the importance of the temporal demand the operator is subjected to, i.e., technical and organizational demands. In particular, he has to plan the normal interventions so that they are performed within allowable time periods and he must control the treatment of the incidents in order to avoid any prejudicial consequence. The major mean he can use is the anticipation which allows him to take advantage of the time tolerances. This anticipation can concern a more or less extended time period (Leplat, 1980) (2) and thus can result in setting a planning of more or less longer range. The mechanism of this anticipation is, by nature, not observable and can be known only through its outward signs or symptoms. So we will try to trace them. A more complete analysis would have used systematically the oral reports of the operators but in the conditions of the study this could be done only very partially.

We attempted to define some symptoms of anticipations. For example in the process scheduling several symptoms can be used such as the early execution of an operation (almost as soon as this execution is possible), the frequent consultation of an indicator which allows the operator to detect the trend of variations and thus to be able to act on them in order to regulate the process according to future activities on other processes. Many symptoms have to be interpreted according to the context and to information which verbalizing will bring.

Informal communications constitute another type of symptom of anticipation. These communications are those the operator decides to make for example to obtain information concerning the times at which departments before and after him in the process intend to intervene. These connexions allow him to anticipate through the widening of the field of variables taken into account.

5- Planning the interventions: coïncidences and priorities. The planification becomes difficult in the case where two (or more) interventions have to be performed at the same moment. A situation of conflict (or coïncidence) occurs thus and the operator is led to implement, implicitly or explicitly some priority criteria and to plan his activity. We will first indicate how such situations can be spotted and then which priority criteria can be used.

Spotting of simples cases of coïncidence. A typology of the various cases of coïncidence could be established and diversified namely by considering the situation resulting from the operation of several fermenters (coïncidence or normal interventions to be carried out on several processes; of normal interventions to be carried out on several processes; of normal interventions and interventions related to incident recovery and including more or less fine distinctions). It would be also possible to conceive a computer-aided treatment of some data, especially for the processes and thus to simulate the effect of incidents the occurrence of which would be defined arbitrarily.

Example of coïncidence: the operator begins to prepare the additive solution 4 hours prior to the transfer because he is aware that one hour later a coïncidence will occur with the heating of a fermenter, the inoculation of a second fermenter and the preparation of this additive solution (comment: "I do it now because I have a little time. Later I would have to deal with the two fermenters".

The criteria for priority: when coïncidence cases occur, some criteria for priority should be used to define the rules governing the organization of the activity.

- Priority should be given to the phases having a strictly limited duration (i.e., with no tolerance) e.g., the sterilization of the medium.
- Priority should be given to the phases having time tolerances which cannot be exceeded without severe consequences (e.g., the inoculation time, if exceeded, could result in irreversible consequences on the duration of the culture and thus, on the planning set for the following week and usually established in advance by the foreman).

The determination of priorities is not always so simple because it may require the consideration of organizational criteria (e.g., organization rules for the shifts of upward and downward services) or a good technical knowledge of the processes (e.g., to determine the consequences of possible mistakes or to appreciate the importance of a preventive control).

## CONCLUSION

This study revealed a few characteristics of the activity involved in the control of a group of processes running in a non-synchronized manner. It suggests some questions about the analysis of activities of this type. To progress and to verify more precise hypotheses it would have been necessary to have more time and to collect a sufficient number of observations, but this was not possible in the conditions of this study. As a conclusion we will draw some ergonomic consequences from the previous analyses.

1- Adjustment of time-related working conditions. It should be said more exactly: adjustment of external working conditions affecting the time-related demands.

a- Reduction of movements by grouping the fermenters together (this is under progress); by installing a control console in the workshop (now under study). It would display information concerning the whole of the fermenters and would permit to avoid a lot of movements, either toward the control room or toward the fermenters during the automatic phase (during which the work is limited usually to the checking for good processing, without any intervention).

b- Improvement of planning. The question here is to improve the precision of the estimations of the duration of the different phases and of the time intervals. To this end, it should be required to store the collected duration data and to perform a statistical analysis to take benefit of the past information to set a more precise planning. The data processor could be used subsequently for establishing these plannings and for updating them depending upon the hazards. The elaboration of a more precise planning should not result in impovershing the task of the operator who needs some motivation to play his role for compensating the deficiencies of the planning. In this line it would be possible to improve the qualification of the operators in order to make them able to use the data processor for introducing the required data (a task performed at the moment by the technicians) as well as for better taking advantage of the computer capabilities. There would be here a way for job enrichment and a means of maintaining some autonomy.

c- To reduce the frequency of the incidents. To this end should be established and updated at regular time intervals a list of the incidents mentioning the involved parameters: frequencies, group, recovery time, etc. The preventive action could be then initiated selectively. It appeared that the incident related to foam development with possible overflowing was particularly frequent. Such an incident could have been reduced by improving the anti-foaming agent; it will be also reduced by improving the foam-level detecting sensor so that it is activated only at the critical moment and not untimely where the foam is of little density and could disappear spontaneously. An improvement of the automatisms could be also looked for so that a wider range of deviations could be taken into account.

d- Improvement of the reliability of the facilities especially those designed for the preparation of the additive solution.

2- Improvement of the anticipation capacities. It has been shown how anticipation is important for a better work planning extending over a longer period of time. Several actions can be taken to confer a wider anticipation capacity to the operators.

a- Improvement of the knowledge of the system so that the operator having a good idea of the operation could anticipate the effect of some situations and of some of his interventions. Such a knowledge could be given to him in an operational way, i.e., under consideration of the actions he has to perform.

b- Improvement of the knowledge of the incidents. The importance of this knowledge has been often highlighted (Salvendy and Seymour, 1973) (7). A listing of the incidents could be prepared with mentions for each of them of how it can be detected (symptoms), anticipated (warning signs), corrected (treatment or recovery of the incident) and prevented. The numerous works dealing with this problem in other fields could be usefully considered here (Rouse and Rasmussen, 1981) (5).

c- Training in evaluation of priorities. In time-sharing-based works, like this one, the operator has to evaluate situations and to make a decision to define an order of performance. It could be possible to train the operator to evaluate adequately these priorities by submitting some cases -by means of written documents, of sliders, of simulation- and asking him to decide the order of the interventions. Such static or dynamic simulations could be accompanied by questionnaires or interviews aimed at detecting and eliminating the origin of the operators' deficiencies.

d- Training in data processor use. The data processor could be used for training and improving the operators. Little simple computer-aided teaching programs could be developed. It would be of highest interest for the operators also to become skilled in computer use not only to get simple information but also to be able to exploit the processing possibilities. No doubt it would be useful to associate them to the exploitation of the data or a least to keep them informed thereon. These suggestions do not exhaust the contribution ergonomics is able to bring: they are only examples and their development requires more deeply conducted analyses. It has been seen however that even a basic analysis is likely to suggest numerous measures of adjustment.

## REFERENCES


1. Conrad, R., 1954, "Speed stress", in W.F. Floyd and A.T. Welford, Human Factors in Equipment, Lewis, chap. 13.

2. Leplat, J., 1981, "Task analysis and activity analysis in situation to field diagnosis", in J. Rasmussen and W.R. Rouse (Eds), Nato conference series, New York, 287-300.

3. Leplat, J., and Savoyant, A., 1984, Ordonnancement et coordination des activités dans les travaux individuels et collectifs, à paraître dans le Bulletin de Psychologie.

4. Mackworth, J.F., and Mackworth, N.H., 1956, "The overlapping of signal for decisions," American Journal of Psychology 69, 28-47.

5. Rasmussen, J., and Rouse, W.R., (Eds), 1981, "Human detection and diagnosis of system failures," Nato conference series, New York, 716 p.

6. Rocher, M., 1982, "Les activités de contrôle de processus partiellement automatisés dans une installation de fabrications biochimiques," Mémoire pour le Diplôme d'Ergonomiste C.N.A.M., 180 p.

7. Salvendy, G., and Seymour, W.D., 1973, Prediction and development of industrial work performance, J. Wiley, New York, 351 p.

8. Theureau, J., 1981, "Eléments d'analyse temporelle du travail infirmier," Le Travail Humain, 93-108.

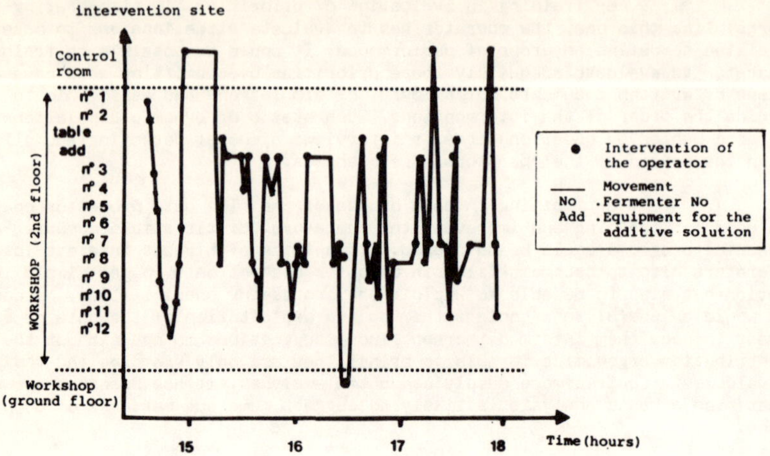


Figure 1 DIAGRAM OF OPERATOR MOVEMENTS

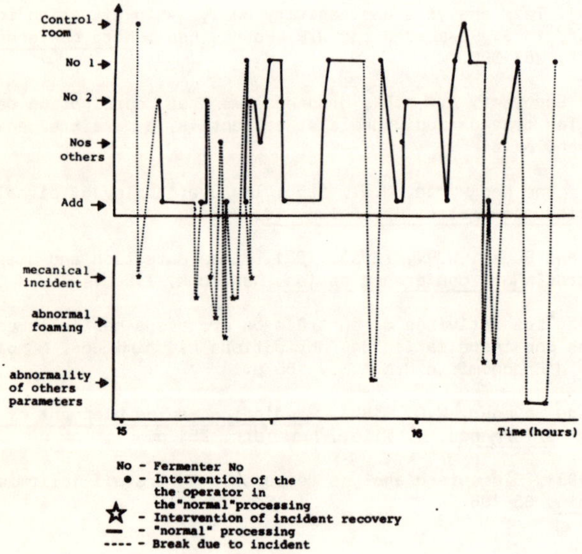


Figure 2 Breaks in the "normal"processing of the activity (upper part),by incidents (lover part)

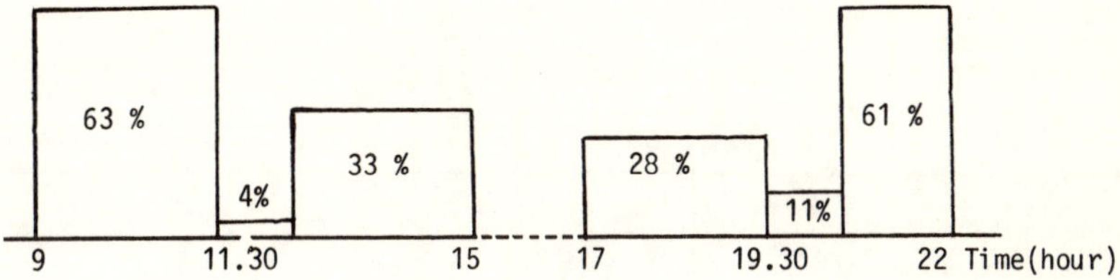


Figure 3 Deviation of real activity hours from planned hours. For each time interval, the frequency of the observations is mentioned on the last line ; e.g.,63% means that in 63% of the cases the reception of medium has been performed earlier.

# SOME RESULTS ON OPERATOR PERFORMANCE IN EMERGENCY EVENTS

D. D. Woods*

This paper synthesizes results from several studies of operator performance in simulated power plant emergencies and from retrospective analyses of operator performance in five actual power plant critical incidents. This synthesis is feasible because all of these studies used process tracing techniques and a common perspective on decision making as the basis for their analysis of operator performance. The paper focuses on two areas: operator's ability to detect and correct errors and how operators utilize procedures. The results in these areas are assessed in terms of their implications for concepts and models relevant to operator performance and in terms of their implications for man-machine system improvements.

## INTRODUCTION

Understanding how operators perform during emergency events is critical to the design and evaluation of person-machine systems in process control applications. This paper examines a data base on operator performance in nuclear power plant emergencies with respect to two issues[1]: what is the operator/crew's ability to detect and correct their errors and how are procedures utilized during emergencies? The data on these questions are assessed in terms of their implications for models of operator performance and for person-machine system improvements.

A data base on operator performance in emergencies now exists through a series of studies of simulated and actual nuclear power plant critical incidents. The data base includes:

a. Woods, Wise & Hanes (1) -- a study of the performance of experienced operators in simulated emergency events (including multiple failure events) with and without a computerized operator aid (39 scenarios from 8 crews in 7 events);

b. Woods & Roth (2) -- a study of the performance of experienced operators in simulated emergency events (15 crews in 4 events);

c. Pew, Miller & Feehrer (3) -- a retrospective analysis of operator decision-making in four actual power plant critical incidents;

*Westinghouse Research & Development Center, Pittsburgh, PA 15235 USA

[1]The results described in this paper represent a portion of a more complete assessment of this data base.

d. Woods (4) -- a retrospective analysis of operator decision-making in one actual power plant emergency event (an additional facility to support emergency operations, Technical Support Center, was active during this incident).

A synthesis of results across these studies is feasible because they all used a process tracing methodology and a common perspective on decision-making [cf., Rasmussen (5)] to analyze operator performance. Operator behavior was analyzed with respect to categories of monitoring plant data, interpretation of this data (knowledge or belief state, intention, expectation, recovery strategy), the resulting control actions, and feedback on the results of recovery actions. These concepts were used to chart, not just what actions the crews took, but also the decision process and context that led to the actions. Figure 1 is an example of the flowchart format used in (1) and (4) to summarize the results of the decision process analysis. In each study, critical decisions identified in these protocols were subjected to a more detailed analysis based on the particular objectives of that study. The results reported here are based on a review of the decision making data base represented by this set of studies.

## OPERATOR'S ABILITY TO DETECT AND CORRECT ERRORS

### Results on the Detection and Correction of Operational Problems

In general, the operators in these studies performed well. Recovery actions were prompt, there were relatively few operational problems, and those that did occur were usually of little consequence to the final outcome. With respect to the operational problems that did occur, the crews rarely had any difficulties with the initial state identification following the onset of the emergency. In general, when operational problems did occur, it was because operator state identification and control activities gradually became decoupled from actual process state as a function of execution problems or the unexpected. This result suggests an analysis of what happened following the occurrence of some operational problem.

The data in Tables 1 and 2 show that when a crew misidentified plant state or had action execution difficulties, they generally failed to correct their understanding of plant state or to identify execution problems within the duration of the test events[2]. For those problems that were corrected, a relatively long time period elapsed: 4 of the 10 corrected execution problems in Table 1 were detected almost immediately (average of 1:15) but the other 6 took an average of 6:46). Some operational problems were corrected through the intervention of some external agent -- instructor hints in the simulator exercises or, in actual incidents, when a fresh viewpoint entered the situation.

### Implications and Countermeasures

The pattern of results illustrates the importance of concept-driven behavior in operator performance [cf., Woods (6) for a more detailed discussion of concept driven behavior]. The data operators sample and, in particular, how this data is integrated/interpreted depend on the operator's assessment of the situation [cf., Figure 2 for an example of a decision flowchart developed to highlight how internal knowledge structures are activated and how they guide

[2] The execution problems can be considered identification problems in that here we are not concerned with the initial source of the execution problem but with the operator's failure or success in assessing the effect of the action or non-action on process state and goals.

TABLE 1 -- Operator performance: Error correction in simulated emergencies.*

| | No Correction | Corrected Only Through External Agent | Corrected |
|---|---|---|---|
| Problem In State Identification | 14 | 5 | 0 |
| Problem In Execution | 10 | 0 | 10 |
| Total | 24 | 5 | 10 |

*Data on operator performance in simulated emergencies from (1) and (2). These data account for 70% of all of the operator problems in these studies, the remaining operator problems were responses judged as too slow or that were noncorrectable. Given the number of test scenarios (99 using 23 experienced crews in eight different events), there were very few operator problems and those that did occur were of little consequence to the final outcome.

observation and action]. When there is some mismatch between actual and perceived process state, it is very difficult for those involved to modify their view given today's control rooms. Misperceptions of process state are often corrected only when a fresh viewpoint enters the situation. This pattern can be described as a type of fixation or perseveration effect, in that, an assessment of plant state tends to persist independent of supporting data. The crew's problems in the Three Mile Island accident could also be described as the result of a fixation effect, and some of the recommendations/requirements to prevent re-occurrence of such accidents can be seen as classic methods to prevent or reduce this type of decision problem (the requirement for a Safety Parameter Display System to provide a concise statement of plant safety status, i.e., provide improved feedback about goal achievement; the requirement for a Shift Technical Advisor can be seen as a way to institutionalize a fresh viewpoint).

The observation of fixation effects in operator state assessment indicates that operational problems are often related to poor feedback about the effect of control actions on process state and on goal achievement. One of the studies in the operator decision making data base (1) examined how concepts for computer-based displays (i.e., safety parameter display systems) might influence operator performance. Results on prototype display system usage suggest that computer-based aids can reduce fixation effects by providing improved feedback especially to senior staff members acting in an emergency or process management role. Table 3 contains examples where feedback was obtained from a computer-based display to correct operational problems.

The error correction results are consistent with the Bartlett (7) multi-level view of cognitive processing [cf., (5) and Broadbent (8) for particular examples of this class of models]. The key elements of this view are:

- o Human processing of information is organized at different levels which operate simultaneously over different time spans.
- o Some (presumably "higher") levels modify or control the operations of others.
- o Lower levels are capable of independent function.

TABLE 2 -- Operator performance: Error correction in actual incidents.

| Study | Decision | Brief Description |
|---|---|---|
| Pew et al (3) | NA1 (pp.B-35 and B-41) | Failure to detect execution problem: Shift supervisor ordered Main Steam Isolation Valve closure, but operator was interrupted before action was taken; shift supervisor was unaware action was not taken. |
| | NA2 (pp.B-37 and B-45 to 46) | Failure to detect execution problem: execution error during realignment to normal charging and letdown; 19 minutes were available to detect and correct error. |
| | OY1 (pp. B-17) | Delayed detection of execution problem: serendipitous detection and correction of misaligned valves by external agent 33 minutes after error occurred. |
| | OY2 (pp. B-18 to 19, B-22, and B-28 to 34) | Failure to correct state identification problem: due to above (OY1) execution error, there were conflicting measures of reactor water level and no natural circulation; crew was unable to explain conflicting measures, unable to accurately assess inventory status, and erroneously assumed natural circulation was occurring until serendipitous discovery of misaligned valves by external agent (31 minutes elapsed). |
| Woods (4) | 1.C.2 (pp.3-5 to 3-6) | Failure to correct state identification problem: Crew and Technical Support Center failed to re-evaluate an earlier action in light of later changes in plant state; as a result, a potential safety function threat was not detected. |
| | 3 (pp. 3-9, 3-11 to 3-13) | Failure to correct state identification problem: failure to detect and correct conditions leading to pressurizer relief tank rupture into containment (19 minutes available). |

TABLE 3 -- Examples where the feedback used to correct operational problems was obtained from a Safety Parameter Display prototype [Results from Woods et al (1)].

*In event TR3-H the reactor operator detected that the faulted steam generator level was within the wide range instrumentation from Safety Panel B plant status and wide range iconic displays. The balance of plant operator had reported earlier that the faulted steam generator was empty by misreading narrow for wide range level from the control board.

*The senior reactor operator (TR2-H) detected low steam generator levels in two unaffected steam generators from Safety Panel B plant status display. (The balance of plant operator had been slow in re-establishing auxiliary feedwater flow to the unaffected steam generators after stopping all auxiliary feedwater to aid in the steam generator tube rupture diagnosis.) The senior reactor operator then directed the balance of plant operator to increase auxiliary feedwater flow to the unaffected steam generators.

*Safety Panel B plant status display helped the senior reactor operator detect that auxiliary feedwater had not been isolated completely from the faulted steam generator (TR2-H). The faulted steam generator had been isolated, but the balance of plant operator turned on the turbine driven auxiliary feedwater pump to increase unaffected steam generator levels. However, auxiliary feedwater flow also began to the faulted steam generator.

*In event TR2-H the senor reactor operator detected from the Safety Panel B plant status display that only 2 of the 3 unaffected steam generators were being used to cool the primary system. He directed the balance of plant operators to open the third steam generator power operated relief valve.

*In another event (FW3-G), a crew detected that pressurizer level was low and decreasing. The crew isolated letdown and then consulted Safety Panel B wide range iconic for feedback. The iconic display showed the crew that the pressurizer level decrease halted.

The result is that there is a multi-level (not necessarily hierarchical) architecture of cognitive processes. Lower level processes, which might be characterized as more ballistic or parameter driven behavior modes, bring efficiencies to behavior (e.g., stereotypical jumps in Rasmussen's step ladder model) but operate only over a limited range of situations. This tradeoff is balanced by other layers of processing, in the multi-level view, which monitor and modify the application of the lower level routines with respect to error correction (are skilled routines proceeding correctly), goal achievement/setting (are skilled routines achieving the desired effects), and responses to novel situations (which includes the ability to distinguish novel from stereotypical situations). If only the upper levels were available, performance would be slow, awkward and demanding as when we learn a new skill. On the other hand, if only lower levels were available, performance would be effective only when the novel or unexpected do not occur. Cognitive architectures that consist of multiple levels combine the advantages of these two kinds of processing while minimizing the disadvantages. The multi-level view is related to Poulton's distinction between open and closed skills:[3] Open skills employ a relatively variable sequence of components and require extensive feedback from the environment in order to adapt behavior to unpredictable constraints or

[3]Poulton's terminology does not correspond to the control engineering terms of open and closed loops.

disturbances; closed skills employ a relatively static sequence of components, operate with little or highly stereotypical feedback from the environment, and therefore function successfully only in highly regular or predictable domains.

The error correction data show that many operational problems are related to a lack of feedback on the success or non-success of recovery actions. From the multi-level view, this is due to lower level, closed routines running on in the absence of higher level control based on monitoring changes in process state and goal achievement. One avenue then to design more error corrective person-machine systems is to model the control room functional architecture [i.e., the person-machine cognitive system; cf., Hollnagel & Woods (10)] on the multi-level structure of internal cognitive processes. This establishes, in effect, a criterion of cognitive diversity, that is, a philosophy of defense-in-depth with respect to cognitive behavior.

The error correction data also show no sign of a speed-accuracy tradeoff, i.e., longer response times were not correlated with higher quality responses. Similarly, Beare et al (11) found no relationship between speed and accuracy for nuclear power plant operator performance in emergency events, and Wohl (12) found two performance regions for maintenance diagnosis in aerospace applications (performance improved with time only up to some critical time). These results show that, for major operator tasks in emergency events, the quality of operator performance is not time dependent (cf., Woods et al (13) for more detailed discussion of the relationship between speed and accuracy in operator tasks).

## PROCEDURE UTILIZATION DURING EMERGENCIES

### Static Procedures - Dynamic Events

Procedures (especially procedures embodied in a paper medium) are a static, sequential list of activities. However, the data base shows that actual operations are dynamic: events occur at indeterminate times, operations can occur in parallel, items may need continuous or semi-continuous monitoring.

Continuous control tasks are one area where this mismatch is acute. In these tasks a continuously variable output parameter is to follow some target trajectory or avoid some dynamic limit[4]; in discrete control, the operator places a component or system into one of several discrete states (e.g., align a series of valves to provide a flow path from a source to a destination). For the latter class of tasks, traditional procedure can work well, in principle, because there is a relatively fixed sequence of actions to be performed[5]. However in continuous tasks, there is a dynamic component to skilled performance based on manipulating inputs contingent on output-goal relations; tuning inputs to achieve output goals is a skill that defies complete expression as a fixed sequence of discrete actions. This dynamic component is particularly strong in a type of task that occurred frequently in this data base -- a task where several interacting continuous control tasks must be balanced (e.g., the cooldown and depressurization tasks in steam generator tube rupture events on nuclear power plants).

---

[4]Note that these tasks are essentially continuous even if the controls available to the operator (the inputs) are discrete and are adjusted intermittently.

[5]Note that skilled performance at discrete control tasks falls towards the closed end of Poulton's open-closed dimension, in that a basically fixed sequence is followed and feedback requirements are local and stereotypical.

Review of the data base with respect to continuous control tasks reveals that two types of operational problems result from the static procedure/dynamic event mismatch. On one hand, many crews tried to perform continuous control tasks in a more open skilled fashion, but operational problems occurred when the crew did not have complete guidance or knowledge of the constraints and goals that should govern process tuning. On the other hand, crews often performed only a discrete control version of the continuous control task, but here operational problems occurred when there was some disturbance that required flexibility in the deployment of the component actions available to achieve task goals.

To take one case of the latter from the simulated emergency trials (2), a crew violated an outcome measure of performance (maintain greater than $50^{o}$ subcooling -- a dynamic limit) in one continuous control task (primary system cooldown in a steam generator tube rupture event). The crew rotely followed and correctly executed the sequence of procedure steps relevant to this task. However, the goal violation occurred because the crew failed to monitor the goal of the task (subcooling), so that, when the event later deviated from expected course and the crew was further along in the rule sequence, the crew failed to detect and correct the deviation over a period of 14 minutes.

The static procedure/dynamic event mismatch is also relevant to another aspect of procedure utilization. Error checks are often annotated to particular steps in a procedure in order to make sure that operators are using the correct procedure or subprocedure out of the total set available. Operational problems such as those described in the previous example are often treated in procedure space by placing instructions to perform error checks downstream in the rule sequence.[6] While there are specific places in action sequences that suggest or demand such checks (e.g., permissives such as do not begin primary system cooldown until the faulted steam generator is isolated), the data base shows that the dynamic aspects of emergency events allow for the possibility that the symptom that activates the check may not occur until after the crew has passed the particular step containing the rule [e.g., Meyer (14)]. As a result, these error checks also represent a higher level task of semi-continuous monitoring of output measure - goal state relationships that requires support from the interface system (e.g., training, guidance, data, and data organization).

## Goal Oriented Versus Goal Controlled Procedure Usage

A goal oriented system is designed to achieve a goal but operates without explicit use of goal information. A goal controlled system explicitly uses goal information in order to adapt its behavior to changing circumstances in order to achieve goals (5).

It is often assumed that operators do or should rotely follow the sequence of choice and action rules contained in procedures. This philosophy of "take care of the steps and the goals will take care of themselves" represents a goal oriented approach towards procedure usage. However, the data base review shows that emergency operations often demand greater response flexibility, where procedural information about how goals can be achieved is a resource to improve

---

[6] Note the significant secondary design problems that arise in the development of complete coverage: identify steps that tell operators when to go back and monitor output goals, place these steps at appropriate points downstream in the action sequence, test placement of these steps in dynamic trials to show that, given the time operators take to complete tasks, typical or important disturbances will be caught in the error check net.

goal controlled operator performance (Figure 2 contains a decision flowchart of operator performance in a simulated emergency where goal controlled procedure usage occurred, i.e., the crew treated major recovery tasks as goals which guided their detailed observations and actions).

This finding is supported by the error correction and continuous control task results discussed earlier, and it is supported by problems that occurred in operator assessment of the relevance of the currently or nominally active rule set to actual plant conditions. For example, how does a crew decide if a procedure or subprocedure is still relevant after a second failure occurs? In two cases in actual emergencies (2; 14), the crews involved thought that a second failure invalidated the procedure they had been using. In another actual event [Brown & Wyrick, (15) pages 3-21 to 3-52], a crew correctly began to follow one procedure; but later, when the leak increased in size, they failed to activate either the newly relevant procedure or all of the relevant recovery goals. In other cases, operators sometimes successfully adapted and sometimes failed to adapt a procedure or rule to unexpected variations in plant behavior [e.g., NRC (16)]. In addition, there were examples in the data base where rote procedure execution broke down once an operator error occurred. In one case (2), the original error invalidated subsequent steps; the crew's attempts to continue to follow the procedure led to further erroneous actions. These results illustrate the two kinds of operational problems that are the consequence of the mismatch between goal oriented procedures and the goal controlled requirements of emergency operations: (a) problems where rote rule following persists in the face of changing circumstances that demand adaptable responses and (b) problems where goal controlled performance is attempted without the knowledge or guidance to successfully manage resources to meet recovery goals.

The results on operator procedure utilization in emergencies show that rotely followed procedures are inherently "brittle," in that, they are not readily able to handle novel situations, adapting routines to special conditions or contexts, underspecified instructions (e.g., "stop all unnecessary safety injection pumps"), and recovery from errors (cf., Suchman (17) and Brown, Moran, and Williams (18) for more discussion of the semantics of procedures and similar results on instruction following from other work domains). Good operations requires more than rote rule following; it requires the operator to "understand how the various steps of a procedure work together to produce intended effects" (18, p.2).

## Countermeasures

The procedure utilization results demonstrate the need for guidance and tools to help operators meet the goal controlled performance demands of emergency operations. For example, foldout sheets can be used to specify important operator activities which could be triggered at indeterminate times during some procedure and items that need continuous or periodic monitoring during that segment of operation (cf., new nuclear power plant procedures such as the Westinghouse Owners Group Emergency Response Guidelines). Procedures can be developed to support periodic monitoring of recovery goals (e.g., safety function status monitoring through decision trees as in the Westinghouse Owners Group Emergency Response Guidelines or periodic monitoring of an event diagnosis/rediagnosis procedure).

Computerized procedures can also aid continuous monitoring and control tasks. Computer displays can lessen paper management problems with current procedures by showing items that should be monitored at any stage in the recovery process. The advantage over the foldout page is additional flexibility, the ability to show actual values as well as criterion values, and the ability to alert crew members when boundaries are approached or crossed. In addition, operator monitoring and goal achievement assessment tasks can be supported through new computer display systems which present data about the relationship between output measures and process goals. In general, techniques that identify and embody envelopes of performance are an effective means to keep operators from losing sight of the forest when enmeshed in the detailed steps of various procedures.

## DISCUSSION

The data base review with respect to issues of error correction and procedure utilization indicate that operator performance is well described by a multi-level view of human cognitive processing: goal controlled, open skill executive processes monitor and control the deployment of semi-autonomous resources which are more goal oriented, closed skills. This class of models emphasizes that all human behavior is knowledge (or concept) driven; what varies with layer of processing is the type of knowledge that drives behavior: knowledge embedded in a motor program, knowledge built into automatic mental skills (e.g., the heuristics of figure/ground organization and other forms of tacit knowledge), knowledge in the form of explicit rules (derived, learned, or written), knowledge in the form of conscious reasoning. In addition, the multi-level view (and the specific results) points out that the criterion for skilled operator performance is the ability to adapt behavior to changing circumstances in the pursuit of a goal.

There is a mismatch between the model of the operator implicit in past process control interface systems and the above model of operator performance. As a result and as the data base review reveals, two general kinds of operational problems occur. In one case, lower level routines are misapplied or run on in the absence of higher level control based on goal monitoring. In the second case, operators attempt to adapt component skills or maneuvers to achieve perceived goals, but problems occur due to incomplete knowledge of the relevant goals and constraints. The implication is that operator performance can be improved through an operational system (training, guidance, information) that is explicitly based on the multi-level view. New process control interfaces are being developed, tested, and implemented that begin to accomplish this goal; examples include new procedures that incorporate guidance on goal monitoring, computerized procedures, and goal directed information systems/decision aids.

## REFERENCES

1. Woods, D. D., Wise, J. A. and Hanes, L. F. (1982). Evaluation of Safety Parameter Display Concepts. Palo Alto, California: Electric Power Research Institute, NP-2239.

2. Woods, D. D. and Roth, E. (1983). Operator Performance in Simulated Process Control Emergencies. Unpublished Data.

3. Pew, R. W., Miller, D. C. and Feehrer, C. E. (1981). Evaluation of Proposed Control Room Improvements Through Analysis of Critical Operator Decisions. Palo Alto, California: Electric Power Research Institute, NP-1982.

4. Woods, D. D. (1982). Operator Decision Behavior During the Steam Generator Tube Rupture At The Ginna Nuclear Power Station. In W. Brown and R. Wyrick (eds) Analysis of Steam Generator Tube Rupture Events at Oconee and Ginna. Institute of Nuclear Power Operations, 82-030 (Also Westinghouse Research and Development Center Report, 82-1C57-CONRM-R2).

5. Rasmussen, J. (1983). Skills, Rules, and Knowledge; Signals, Signs, and Symbols; and Other Distinctions in Human Performance Models. IEEE Transactions on Systems, Man, and Cybernetics, SMC-12, 257-266.

6. Woods, D. D. (1984). Visual Momentum: A concept to improve the cognitive coupling of person and computer. International Journal of Man-Machine Studies, in press.

7. Bartlett, F. C. (1943). Fatigue following highly skilled work. Royal Society of London, Proceedings, 131, 247-257.

8. Broadbent, D. E. (1977). Levels, hierarchies, and the locus of control. Quarterly Journal of Experimental Psychology, 29, 181-201.

9. Poulton, E. C. (1957). On prediction in skilled movements. Psychological Bulletin, 54, 467-478.

10. Hollnagel, E. and Woods, D. D. (1983). Cognitive Systems Engineering: New wine in new bottles. International Journal of Man-Machine Studies, 18, 583-600.

11. Beare, A., Crowe D., Kozinsky, E., Barks, D. and Haas, P. (1982) Criteria for Safety-related Nuclear Power Plant Operator Actions. NUREG/CR-2534.

12. Wohl, J. G. (1982). Maintainability prediction revisited: Diagnostic behavior, system complexity, and repair time. IEEE Transaction on Systems, Man, and Cybernetics, SMC-12, 241-250.

13. Woods, D. D., Rumancik, J. A. and Hitchler, M. J. (1983). Issues in Cognitive Reliability. Proceedings of ANS Topical Meeting on Anticipated and Abnormal Plant Trancients in Light Water Reactors, in press.

14. Meyer, O. Idaho National Engineering Laboratory, personal communication.

15. Brown, W. and Wyrick, R., Eds. (1982). Analysis of Steam Generator Tube Rupture Events at Oconee and Ginna. Institute of Nuclear Power Operations, 82-030.

16. U.S. Nuclear Regulatory Commission. Information Notice 83-30: Misapplication of generic emergency operating procedures guidelines. May 11, 1983.

17. Suchman, L. A. (1982). Towards a Sociology of Human-Machine Interaction: Pragmatics of Instruction-Following. CIS Working Paper, Palo Alto, CA: Xerox Palo Alto Research Center.

18. Brown, J. S., Moran, T. P. and Williams, M. D. (1982). The Semantics of Procedures CIS Working Paper, Palo Alto, CA: Xerox Palo Alto Research Center.

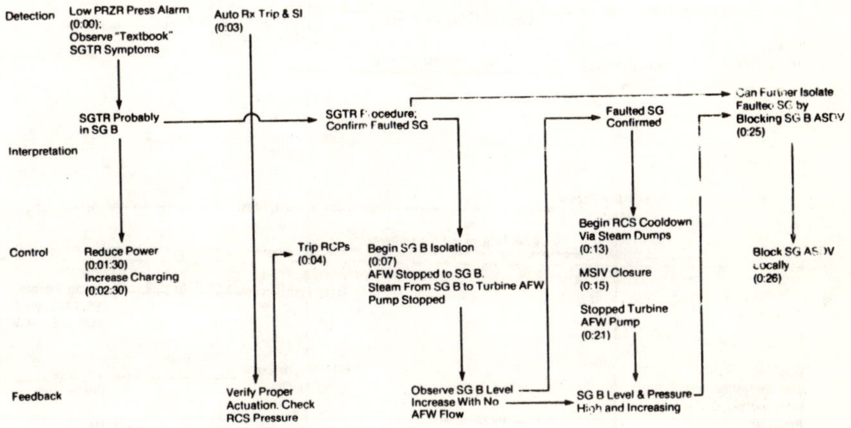


Figure 1 Portion of a decision flowchart from on of the studies in the database (4) that illustrates how decision making concepts were used to understand, not just what actions were performed, but also the decision process and context that led to the actions.

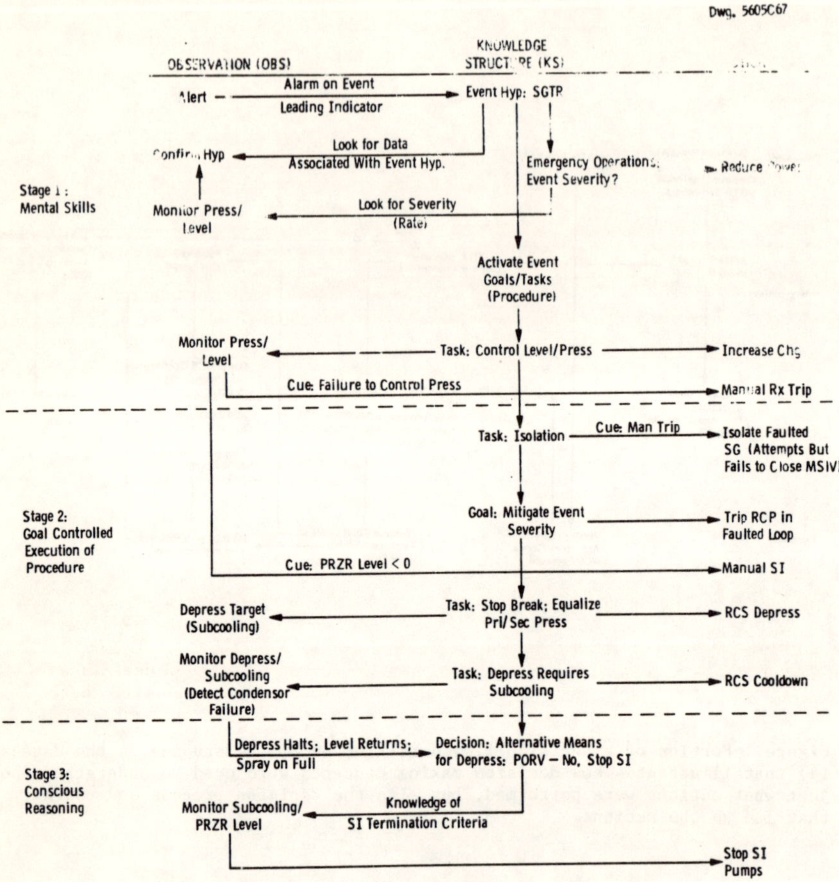


Figure 2. Decision flowchart of crew response in a simulated emergency [steam generator tube rupture from Woods et al (1)]. The format was developed to map how knowledge structures are activated and how they control observation and action. Also note that, during a portion of the recovery, operator performance was based on goal controlled procedure usage.

# WORKSHOP: EFFECTS OF MONOTONY AND REDUCED ACTIVITY IN MONITORING AUTOMATED PLANT

This workshop explores some of the effects which might occur in operators' interaction with highly automated plant. Such operators have to monitor normal operations, and respond appropriately to infrequent incidents and failures. The consequences of such monotony and boredom are likely to be adverse, yet are often unforeseen during system design. Laboratory research on watchkeeping or vigilance tasks has established many characteristics of the human observer's failure to detect critical events, and it is important to consider how these characteristics are reproduced in practical work situations.

## INTRODUCTION

(Convenor: D. Whitfield, Ergonomics Development Unit, University of Aston in Birmingham)

The general aim of this workshop is to gather together and to discuss practical experience and theoretical concepts relating to this problem. With the development of more reliable automated control systems, it seems likely that future process control operators may experience extended periods of relatively continuous and inactive responsibility. They will still be relied on to respond to significant, though fairly infrequent, demands from the system, and this requirement may well be incompatible with the effects of long periods of relative inactivity. There may be a further incompatibility, between the high level of technical experience and qualifications required for adequate response to the infrequent events, and the generally undemanding level of normal operations.

## HUMAN IMPLICATIONS OF TASK AND SYSTEM DESIGN

V. David Hopkin
RAF Institute of Aviation Medicine, Farnborough, Hants

The main reasons for human monitoring of automated processes, and for the consequent monotony and reduced activity, are the following: enhanced system efficiency, productivity or competitiveness; reduced capital outlay or resources such as manpower; technological advances and innovations; greater system reliability. These reasons do not include human factors.

The observed effects on human operators have two common characteristics: they are adverse and unforeseen. They are adverse for three main reasons:

1. The operator may gain the impression that there is no sensible job for the human being to do, that skills are wasted and that system needs are taking precedence over human needs.

2. The residual human roles, for which the operator's presence is still required, cannot be fulfilled effectively because human beings cannot take an active enough part in the control loop to maintain continuously their detailed knowledge of the current state of the system or to maintain their attention.

3. Monotony and boredom may lead to antagonism and even rebellion; people are surprisingly adept at demonstrating the deficiencies of a system they dislike.

The effects are unforeseen for three further reasons:

1. Many of the well documented general consequences of reduced activity and monitoring are ignored during the system design process because there is insufficient informed human factors input.

2. Many of the specific consequences for a particular system (peer pressure, development of professional norms and standards, opportunities to learn and profit from the observed experience of colleagues) are not recognised as pertinent, and it is only when they are no longer there and cannot be restored that their true importance becomes clear.

3. The human consequences cannot be, and are not, included in costing.

For a long time there has been overwhelming evidence that the human being is very poor as a continuous monitor. Every attempt must be made to maintain the operator's attention and involvement as human roles become more passive, but it is unwise to predicate the safety or efficiency of the system on the assumption that such efforts will always succeed (Hopkin, 1). The voluminous laboratory studies on vigilance can also be misleading as the findings are seldom valid in real-life (Kibler, 2). For efficient monitoring, aids (auditory, visual or both) to attract attention when human intervention is required are essential.

The opeatore must have sensible tasks to do (Jordan, 3); it does follow that such tasks must not be done automatically as well. Man and machine do not on the whole fail in the same ways or make the same kinds of error. In principle, maximum efficiency and reliability should therefore be a combination of both, with discrepancies cross checked. Aids should be most useful when most needed - they should not, for example, impose an additional search task only when the operator is busy and has least time to search. Ideally, the operator should have some control over workload, to perform more tasks manually when task loading is light and allow them to be performed automatically under high task loading. A seriously underestimated factor is the extent to which the actual performance of routine manual functions, such as writing, speaking, calling down information, and discarding out-of-date information, as distinct from watching these functions being performed automatically, reinforces understanding and recall of what is happening. It seems probably that this is because the information needed for actual performance has to be processed at a different level from that which is merely monitored. Quite commonly in more automated systems a new human task is introduced whereby the operator calls down information sequentially, not to take any action on it but merely to top up his memory.

Monotony - more usually "boredom" in the literature (Thackray, 4) - is an example of an adverse and unforeseen consequence which in retrospect looks obvious. Relevant data are scattered. Three authoritative reviews of boredom - a general review (Smith, 5), a paper on job design (Guest et al, 6) and a consideration of boredom as a stress (Thackray, 4) - cite generally different literature, and none mentions perhaps the most stimulating textbook on the subject (Csikszentmihalyi, 7). (Perhaps the notion of a stimulating textbook on boredom is a contradiction in terms). Boredom is a subjective state, not a physiological state, a level of performance, a type of task, or a system design characteristic. Purported findings about boredom expressed in such concepts require a strong supporting rationale. Boredom is not inactivity, nor is it repetitiveness. It seems to have something to do with personality, with job expectations, and with social climate: apparently people will not tolerate boredom at work now to the extent that they once did. Boredom can be associated with mastery of a task which has long since lost all its interest and challenge but is performed very well. Boredom is not associated with active learning.

On the whole, the origins of boredom seem to be an interaction between the work itself and of characteristics of individual operators, but the consequences are on attitudes to the system, to the management, to conditions of employment, and to the workspace design, and on attrition and sickness rates, absenteeism and industrial unrest. The alleviation of boredom is not necessarily accompanied by any improvement in system performance, and expectations along these lines are unlikely to be fulfilled. Concepts of work alienation are clearly pertinent (Kanungo, 8). On this whole topic of boredom, there is an alarming tendency to impute causes to mere correlations, and to resort to commonsense in the teeth of the evidence. Of course, people who are bored are likely to fret, to generate spontaneous activity, and to embark on exploratory behaviour - to tinker with items that are best left alone.

References

1. Hopkin, V. D. Human Factors in Air Traffic Control. NATO: AGARDograph No. 275. 1982.

2. Kibler, A. W. The Relevance of Vigilance Research to Aerospace Monitoring Tasks. Human Factors, 7, 2, 93-99. 1965.

3. Jordan, N. Themes in Speculative Psychology. London: Tavistock Publications. 1968.

4. Thackray, R. I. The Stress of Boredom and Monotony: A Consideration of the Evidence. Psychosomatic Med., 43, 2, 165-176. 1981.

5. Smith, R. P. Boredom: A Review. Human Factors, 23, 2, 329-340. 1981.

6. Guest, D., Williams, R., & Dewe, P. Job Design and the Psychology of Boredom. Paper to 19th Internat. Congress of Appl. Psychology, Munich.

7. Csikszentmihalyi, M. Beyond Boredom and Anxiety. London: Jossey Bass. 1975.

8. Kanungo, R. N. (Ed.). Work Alienation and Involvement. Int. Rev. of Appl. Psychol., 30, 1 (Whole Number). 1981.

# PROCESS CONTROL OPERATION AS A WATCHKEEPING TASK

(A. Craig, MRC Perceptual and Cognitive Performance Unit, University of Sussex)

A significant aspect of process control operation resembles watchkeeping. The operator has to be continuously on the look-out for any faults developing in the system, and rare though such occasions may be, he must be ready to take appropriate action when necessary.

But the operator is not infallible. Sometimes he fails to see a fault until it is too late; at other times he may see the fault, but keys in an inappropriate response.

The systematic study of errors arising during prolonged watchkeeping - the problem of maintaining vigilance - has been almost exclusively confined to laboratory research, employing simple, abstract versions of real tasks. The laboratory approach has been justified by demonstrations that the research findings apply to some real tasks in industrial and military situations. The question raised in this contribution to the workshop is whether and to what degree the laboratory description extends to process control operation: what elements of the following description does the practitioner recognize as applying in his particular case?

The laboratory research has been largely concerned with failures to recognize a critical event when it occurs - detection errors - charting their development over the passage of time. Typically, detection failures are more numerous than expected from prior knowledge of the operator's ability; and over a period of continuous monitoring, the likelihood of detection failure steadily increases as a function of time.

Surprisingly, it does not seem to make much difference to error rates whether the operator is continuously active, reacting to each and every piece of information, or only reacting rarely, to the occurrence of critical events; nor, within a moderate range of work-rates, does it matter whether the work is paced, or whether the operator goes at his own rate.

A number of factors that tend to increase the likelihood of detection errors, have been identified. They include: unfamiliarity with the kind of work involved, a low rate of occurrence of the critical events, the occurrence of critical events at improbable display locations, the presence of adverse environmental conditions such as high ambient temperature, lack of sleep, night work, depressant drugs, and an inappropriately high expectancy of critical event rate at the start of the work period. Poor individual performance has been partly identified with personality variables, especially extraversion, and occasionally with age. Low motivation has been expressly cited as a major cause of increased errors over time.

Errors can be reduced, their increase over time diminished by several known procedures. These include: Administering a stimulant drug beforehand; giving feedback about performance during the task (even if the feedback is not accurate); alternating periods of work and rest at half-hour intervals - or inserting a mere 5-minute rest break every half-hour; introducing the presence of a supervisor, even if he never says anything. Performance levels tend to be higher in introverts, particularly in the early morning. Some evidence indicates that experienced observers tend to produce low but stable performance levels.

In general, detection failures tend to be worse at night than during the day. This distinguishes them from the errors of inappropriate responses or of faulty memory that seem to be higher during the day than at night. In addition, the latter types of error seem to be more resistant than detection failures to the increase over time.

The laboratory research has put forward explanations for the failure to maintain vigilance. One account suggests that the action of looking for a critical event and signifying its occurrence gradually weakens because there is no attractive payoff for doing so. Encouragement and feedback about correct detections will strengthen the behaviour, and so reduce the errors; when spurious critical events are introduced for this purpose, performance reliably improves. Another explanation is that the operator gears his reactions to the expectancies he has built up about critical event occurrence; if critical events are rare, then he tends to dismiss all but the most obvious occurrences as being just noise or jitter in the system. The injection of artificial "signals" tends to increase the detection probability for true signals.

Another explanation suggests that vigilance declines because the workplace does not provide a sufficient variety of stimulation to maintain the brain's arousal at a level that is adequate for efficient functioning. Thus, providing music, engaging the operator in occasional, brief, conversation, or getting him to stand and stretch from time to time, have all been shown to reduce detection failures.

Again, the question for the practitioner is: do these recommendations work for you? or, in terms of their practicability - could they work for you?

This outline is intended to serve as the basis for a dialogue. I have sketched the findings and beliefs of the research workers, to draw them to the practitioners' attention. Can they in turn present observations and data to corroborate our findings, and substantiate our beliefs, or to disconfirm them?

## References

1. Craig, A. Human Engineering: the Control of Vigilance. In J. S. Warm (Ed.) Sustained Attention in Human Performance. Chichester: Wiley, 1984.

2. Craig, A. and Colquhoun, W. P. Vigilance: a review. In C. G. Drury and J. G. Fox (Eds.) Human Reliability in Quality Control. London: Taylor & Francis, 1975.

# WORKSHOP: HUMAN RELIABILITY ASSESSMENT TECHNIQUES

This workshop consists of six contributions which discuss practical aspects of human reliability assessment in a variety of contexts. Four of the contributions involve the use of various techniques to quantify the likelihood of human error in various engineered systems. One paper describes typical errors which give rise to fires and the remaining contribution considers the social and organisational aspects of accident causation with particular emphasis on communications failures.

## INTRODUCTION

(Convenor: D.E. Embrey, Human Reliability Associates Ltd., Dalton, Parbold, Lancashire)

The initial impetus for this workshop came from the Human Reliability Assessment Subgroup (HRAS) which is part of a larger working group, the Human Factors in Reliability Group (HRFG). This group consists of representatives from the universities, and companies in the power, chemical processing and other industries who are interested in reducing costs and increasing safety and reliability through the reduction of human error. HRAS has focussed specifically on the assessment aspects of human reliability studies.

The object of the workshop is to provide a cross section of current applied work in the human reliability assessment area, with particular emphasis on case studies and the quantification of human reliability. The first paper by Whitehouse discusses a wide variety of incidents involving industrial fires in which human error has been significant. The following two papers by Hensley and Embrey, discuss various approaches to the quantification of human reliability. Hensley first describes the use of subjective judgement and the TESEO approach to assess the likelihood that a developing control room fault situation would be detected. His paper also describes the application of the widely used Technique for Human Error Rate Prediction (THERP) to the same problem. In the paper by Embrey, another technique employing expert judgement, the Influence Diagram, is applied to assess the likelihood that an operator will carry out a critical action in a nuclear power station emergency. The use of this approach to produce design recommendations is discussed. The next paper by Bellamy describes the ways in which organisational and communication failures can give rise to system failures. Williams and Featherstone describe a generalised approach to human reliability assessment based on event trees and fault trees, which they illustrate by reference to a non-destructive testing application. The final paper, by Webley, provides a detailed discussion of both the qualitative and quantitative aspects of human reliability assessment, in the context of an emergency drill in a nuclear submarine.

## A PARTICULAR FACET OF HUMAN UNRELIABILITY - PEOPLE AS THE CAUSE OF INDUSTRIAL FIRES

(H.B. Whitehouse, Insurance Technical Bureau, London)

It is very easy when looking for the cause of an industrial fire, to look at things - but nearly all are in fact caused by people. The Home Office Statistics for 1978 -81 show that :

28% are caused by deliberate acts
31% are caused by accidental acts
30% are caused by faulty equipment
11% are of unknown cause

for industrial and commercial fires in buildings. The great majority of the "faulty equipment" category, and the "unknown" are, I believe, also people-induced. One can say therefore that roughly 90% are the result of human unreliability.

The following cases have been picked more or less randomly from reports where there was sufficient detail of the cause to be reasonably sure of the chain of causality, to show that even when "things" seem to be the cause, people are, in reality, the cause.

Gas main explosion: An experienced foreman used a match to test if a main was free of gas. (London, 1880).

Flixborough : A temporary by-pass pipe in a pressurised system failed, releasing several tens of tons of flammable material. Was it faulty equipment?

Chemical plant : An internal explosion released about 50t. of flammable material. Why did the control system fail to prevent the explosion?

Chemical plant : An internal explosion was due to an imperfectly performed daily test of the safety system.

Chemical plant : A small-scale batch process went out of control, causing an explosion, due to too much of a reactant being added.

| | |
|---|---|
| Oil refinery | : Mal-operation of a sampling system on a pressurised propane sphere released a flammable cloud : after ignition lack of emergency control allowed escalation. |
| Pipeline | : A buried pipe carrying LPG burst - at about 50% of yield pressure. The line was still in use - after 29 significant failures. |
| Refinery | : A plug cock in a drain system had been incorrectly assembled; the operator thought it was shut when it was in fact open. |
| Packaged boiler | : Incorrect manual start-up (automatic start-up was available) caused a firebox explosion. |
| Deluge system | : Rendered inoperative by failure to close a priming valve. |
| Loom | : A maintenance man (20 years experience) failed to locate and clip a cable correctly : it was struck and the short circuit ignited oily fly. |
| Welding | : Too well known to need details. |
| Three Mile Island | : A classic compounding of errors. Two quite small mechanical problems which were not new, and lots of human ones. |

Many others, large and small, could be quoted - and none of them would have happened if everyone had done what they should have done. In most cases, the people concerned knew what they should have done. In nearly all the others the people directly concerned should have known, if others had done their job properly.

The actual degree of unreliability is small. There are roughly $10^5$ fires per year in industrial or commercial buildings. The working population is roughly $10^7$ so 1 in 100 people are "triggers". The number of opportunities we have per year is speculation - perhaps 1000? The residual error rate is therefore about $10^{-5}$ per year, but the losses are about £400 million per year.

## QUANTIFICATION OF HUMAN RELIABILITY IN A CONTROL ROOM SETTING USING THREE DIFFERENT APPROACHES

(George Hensley, British Nuclear Fuels Ltd., Risley)

### Background

The following scenario is one that was considered as part of a probabilistic risk assessment and serves to illustrate some of the issues addressed through the approach adopted.(1)

In a particular nuclear chemical plant, consideration was given to the frequency at which a specific hazard might arise. Whilst the consequence of such an event would be small, it was desirable that the frequency of the hazard should be very low. The onset of the fault condition would be revealed to the control room operator by audible and visual alarms, by an "alarm" print-out on a desk-mounted data logger and by instrument readings. A considerable amount of time (about five hours) would elapse before a dangerous state would be reached and the operator would be alerted by at least three separate sets of instruments, with redundancy and/or diversity in each set, and by the appropriate alarms. Although the operator's ability and training was such that he could be considered to conform to a skill-based pattern of behaviour, a separate check against this fault condition would be made every two hours by the shift manager. He would be a professionally qualified scientist conforming to a "knowledge-based" behavioural pattern, thus providing a different type of response to that of the control room operator.

## Hardware Aspects

The redundancy and diversity provided in the instrumentation control and alarm equipment was such that the probability of its failure was estimated to be at a level where total failure would be dominated by common mode faults, that is, at about the $10^{-4}$-$10^{-5}$ probability of failure on demand - bearing in mind the relatively long period of time (five hours) required to go from fault initiation to the hazard materialising.

## Human Reliability Aspects

An estimate was also made of the probability of failure of the control room operator and the shift manager. This was done using three different approaches; subjective judgement, TESEO modelling and THERP/TESEO modelling.

Subjective Judgement: The overall probability of failure of the human system was decomposed into the two elements; operator failure and shift manager failure. These values, when combined, gave a total value of $10^{-5}$.

TESEO Modelling: The control room operator was considered as making three independent attempts at correcting a developing fault condition, while the shift manager would have two. These five occasions were assessed by assigning values to the five TESEO factors (2) as appropriate for each occasion, and then combining the five error probabilities using a multiplicative rule. This gave a value for the human system of $<10^{-5}$.

THERP/TESEO modelling: As there was no appropriate value in the human error probability data of Swain and Guttmann (3) the value derived in the TESEO model for the operator's first opportunity to correct the fault condition, was utilised. The THERP model of dependence (Table 10-2 in the Swain and Guttmann Handbook) was then used to compute values for the second and third occasions, given the TESEO value for the first attempt, using appropriate values for the dependence between these attempts. These values were then combined using a multiplicative rule. The most appropriate set of conditions that applied to the shift manager were those of "usual monitoring in NPP with some kind of check-list or procedure - but without written materials" and this value from Swain and Guttmann was used, assuming a suitable value for dependence between the two opportunities, to compute the shift manager's error probability. The error probabilities for the operator and the manager were combined to give the overall value of $10^{-4}$.

## Conclusion

From the use of three different assessment approaches going from a basic estimate to a simple model, then a more complex model, a comparable set of values were obtained for the probability of complete failure of the human system. The highest value of $10^{-4}$ was in agreement with the upper end of the range corresponding to the probability of failure on demand of the control and instrumentation system. This analysis ignores the fact that an improvement in human reliability could be gained where the "operator" had the opportunity to recover from his error. This would have been the case in this instance because of the cues available from the display system.

## References

1) Hensley, G., Keery, R.J. and Moulding, T.L.J., 1983. "The philosophy of control centre design and manning". Symposium on Human Reliability in the Process Control Centre, UMIST, Manchester.

2) Bello, G.C. and Columbari, V., 1980. "The human factors in risk analysis of process plants: the control room operation model TESEO." Reliability Engineering 1.

3) Swain, A.D. and Guttmann, H.E., 1983. "Handbook of human reliability analysis with emphasis on nuclear power plant applications." Final Report.

# THE QUANTIFICATION OF HUMAN RELIABILITY USING INFLUENCE DIAGRAMS

(D.E. Embrey, Human Reliability Associates Ltd, Dalton, Parbold, Lancashire)

## Introduction

The Influence Diagram Approach (IDA) is one of a number of techniques based on quantified expert judgement which can be used to evaluate the probability of human error in systems. These techniques are particularly useful in analysing high risk scenarios for which little or no empirical data are available.

The Influence Diagram is also a very effective way of

'raising the consciousness' of designers with regard to the ways in which different influences such as procedures, training, etc. interact together to affect the likelihood of human error in a particular situation. The application of Influence Diagrams to human reliability evaluation has been developed jointly by L.D. Phillips of the London School of Economics, and Human Reliability Associates Ltd.

## The Influence Diagram

The form of a typical Influence Diagram is shown in Figure 1. It can be seen that the likelihood that a particular human action will succeed or fail is modelled in the situation being considered as being affected by three major influences : the quality of the information available to the operator, organisational influences, and personal factors. These influences are in turn affected by lower order influences. For example, the quality of the information is modelled as being influenced primarily by design factors and the meaningfulness of procedures.

Only the major influences on success or failure are represented explicitly in the Influence Diagram. Other influences will be present and this is taken into account in IDA, although they are not modelled directly. The general form of the Influence Diagram is developed through interaction with the analyst and subject matter experts such as operators and trainers. Detailed definitions of the dimensions constituting the lowest order influences are derived in bi-polar form. For example, the characteristics of good displays are defined in terms of attributes such as: 'easy to read and understand, accessible, alarms discriminable' etc. Poor displays are defined as: 'hard to read, difficult to interpret, inaccessible' etc. The two extremes of the credible range of the characteristics defining each influence are specified in a similar way.

The Influence Diagram can either be derived off line, or in conjunction with the judges actually carrying out the numerical assessments. We have found the Influence Diagram shown in Figure 1 to be reasonably generic. However, it is easy to modify the structure or redefine the influences in a particular application if necessary.

## Using the Influence Diagram to Generate Probability Assessment.

To exercise the technique, a group of expert judges are required. These should include individuals with knowledge of the specific system being assessed, and might comprise operators, designers, PRA analysts etc. After the judges have been thoroughly familiarized with the structure of the Influence Diagram, and have considered and amended the influence definitions as deemed appropriate, they are then asked to generate balance of evidence estimates (BOEE) for the lowest level influences.

A BOEE is a numerical assessment by the judges of the degree to which the influence, in the specific plant and for the specific human action being considered, tends towards the 'good' or the 'bad' extreme, (i.e. enhances or degrades the likelihood of success). Thus, in the case of the lowest level influence 'meaningfulness of procedures' a BOEE of zero would indicate that procedures were as poor as they could credibly be in a real plant, whereas a BOEE of 100 would imply that they were as good as was feasibly possible.

After the judges have generated BOEEs by a consensus process, they are then asked to carry out a similar process for the effects of all combinations of lower level influences on middle level influences. An example would be 'If design is good, and procedures are meaningful then the BOEE that the quality of information is high is X'.

In the final stage of the procedure, the combined effects of each of the middle level influences on the probability of success or failure are estimated. This generates a series of questions of the form: 'If the quality of the information is high and the organisation is effective, and personal factors are favourable, what then would be the probability of success/failure?'

The calculation procedure which derives the overall probability of success/failure from the IDA procedure uses the lower and middle level BOEEs to derive weights, which represent the relative impact of the lower level influences on the top level probability estimates. These are applied to the judges' probability estimates of the success/failure of the required human action given the various combinations of middle level influences, to give the required overall probability.

The procedure outlined above iterates until the judges are in reasonable agreement regarding the result. Sensitivity analyses are are usually carried out on-line to determine the critical influences.

## Conclusion

IDA has been applied to Risk Assessment in both the offshore and nuclear industries. The capability of the technique to generate design recommendations and the detailed insights that it provides into the situation being analysed, has given rise to a high level of user acceptance. Although further research and development is needed, initial experience suggests that this approach has considerable viability for the assessment of human reliability. The use of the technique in a specific nuclear power context will be described in the workshop.

## INVESTIGATION OF LARGE SCALE ACCIDENTS AS A METHOD FOR OBTAINING HUMAN RELIABILITY DATA

(L. Bellamy, Ergonomics Development Unit, University of Aston, Birmingham)

No human reliability quantification technique can be successful in real life situations unless it can take account of the majority of factors that can significantly influence human performance. Sources of data on performance shaping factors (PSFs) can range from carrying out controlled laboratory research to investigating possible causes of incidents of error that have occurred in real life situations. The aim of the present discussion is to illustrate the value of systematically analysing major accident reports. Examples of specific accidents will be given, showing how the analyses can be organised into a coherent framework.

This author has attempted to uncover data concerning the influence of the individual, social and organisational factors on performance. Swain's (1974) extensive list of PSFs, for example, includes these factors as relevant variables. However, a scan of the research literature provides a very limited indication of the specific features of the social or organisational environment that can contribute to error or the types of error that might be associated with them. This does not mean that such features cannot be identified. A particularly rich yet poorly tapped source is the accident report, specifically those reports documenting evidence obtained from lengthy public inquiries into large scale accidents.

A current survey of a variety of accidents by this author has shown that causes and types of error in the social and organisational domain can be readily identified. Not only this, but in most large scale accidents such errors were prominent causes, many reflecting weaknesses in the interpersonal communication process or communication structure often involving large organisational areas. Communication problems can bring about a "mismatch" between an individual's belief about the state of the system and the actual state of the system. The occurrence of mismatch can, for example, result in distorted perceptions, failure to follow through the implications of an action, or simply a failure to diagnose the current system state.

The ways in which individual needs and beliefs are determined by a particular situation are often complex and interrelated. The existence of a weakness in the communication structure can be a necessary but not sufficient cause of an accident. A situational change may precipitate a major accident, the causes of which are based on errors which may gave gone unnoticed for some time simply because before the change they would have had no immediate observable consequences. This author would wish to argue that the incubation period preceding a major accident is not in itself unique but is a distinct pattern that is often repeated. Although complex, this pattern of human error can generally be reduced to a few interrelated major factors. Hopefully, identification of these factors will aid in identifying and eliminating organisational weaknesses and in the quantification of human reliability in real life systems.

1. SWAIN A.D., 1974. The Human Element in Systems Safety. London: Industrial and Commercial Techniques Ltd.

## A METHOD FOR QUANTIFYING THE EFFECTIVENESS OF ULTRASONIC INSPECTION TO IDENTIFY AND LOCATE FLAWS IN WELDS

(J.C. Williams and A.M. Featherstone, United Kingdom Atomic Energy Authority, Safety and Reliability Directorate, Culcheth, Warrington.)

This paper combines the technologies of human factors engineering and reliability assessment to propose a generalised methodology for quantifying the effectiveness of ultrasonic inspection to identify and locate flaws of a given size in a given weld using a specified technique. To take account of the time sequencing of events in weld inspection it is proposed that an "Event Tree" be used. Emanating from the Event Tree are "Fault Trees" which model and quantify the effectiveness of the ultrasonic inspection.

The "Fault Tree" in this case is the generic model of the ultrasonic inspection; it is used to describe how the various processes involved in ultrasonic inspection can lead to failures of identification and location. The mid-term events in the fault trees are generic terms such as failure to scan correctly. This term, and its associates, are then broken down, finally, to their initiating events, which again are generic terms. The methodology has been designed for the assessment of a specific example. However by only using generic terms throughout it ensures that the method remains applicable to a wide variety of problems.

Ultrasonic inspection is a non-destructive testing technique which, of necessity, requires a considerable amount of skill and judgement by all concerned. Human factors analyses (sensitive to the effects of skill and judgement) in combination with reliability technology, have been found to be a powerful method for generalised assessment of this very important industrial procedure.

## THE USE OF A 'THERP' APPROACH TO GENERATE HUMAN RELIABILITY ESTIMATES FOR A PROBABILISTIC RISK ASSESSMENT

(K.M. Webley, Risk Assessment and Safety Studies Group, Rolls-Royce and Associates Ltd, Derby)

This case study describes how using the 'Technique for Human Error Rate Prediction' (THERP) approach, estimates of operator reliability required for a Probabilistic Risk Assessment carried out using an Event Tree - Fault Tree methodology have been generated.

The first part of this case study describes a category of loss of coolant accident which can occur on a Pressurised Water Reactor the required operator response in the early stages of the accident, and the event tree used in the probabilistic Risk Assessment to model the early stages of the accident.

The second part of the case study shows how the Human Reliability Estimates required to evaluate the event tree were derived. The way the THERP approach was used to synthesise the required values is described. Included in this is a description of the basic human reliability data and performance models required for the evaluation.

The remainder of the case study analyses the strengths and weaknesses of a THERP approach and illustrates how the approach can be used to identify the ways of reducing the frequency of undesired consequences.

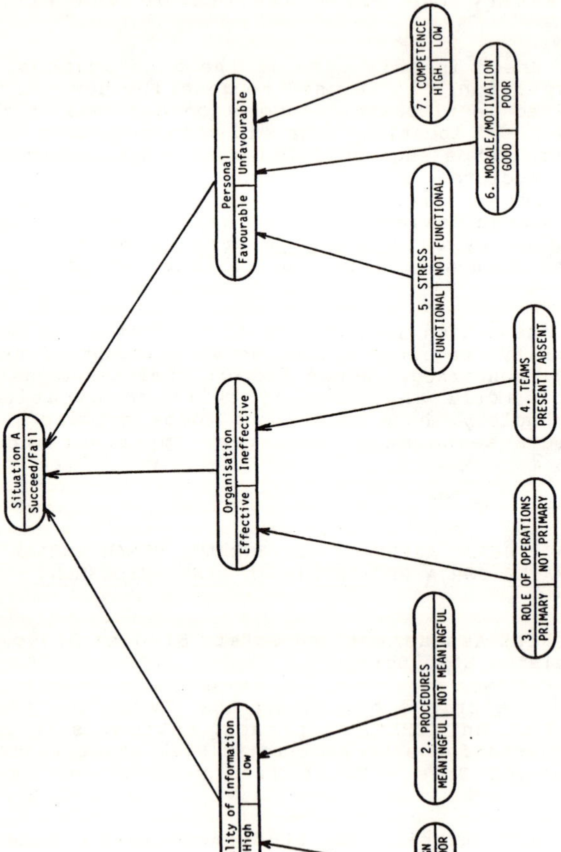


Figure 1 A Typical Influence Diagram

# EXPERT SYSTEMS IN PROCESS PLANT FAULT DIAGNOSIS

P. K. Andow *

All modern process plants of any significant size are equipped with automatic control and protective systems. A particular problem concerning the human operator is his task of analysing alarms to diagnose the basic causes of disturbances. Early and accurate diagnosis is needed if failures are to be detected (and perhaps corrected) before trip systems shut the plant down.

Expert systems provide a possible method of aiding the operator with the diagnostic problem. An experimental system is described which attempts to help the operator make effective use of the alarm system and other displays.

## INTRODUCTION

Fault diagnosis is an important component of the operators job. Two major trends are easily observed that are most relevant to this task:

1. Plants are increasingly designed to be operated from central control rooms.

2. There has been an ever-increasing use of automatic controls and protective equipment.

When coupled together these trends have resulted in small operating teams being able to monitor and control large plants. Often the operators cannot even see the plant, parts of which may be physically remote from the control room. (In the case of distribution or communication control systems it may be that all of the plant is remote). There has also been an inevitable tendency to automate those control functions which are easy to understand and engineer. The operator is then left with a heterogeneous collection of tasks which are not easy to understand or engineer.

The type of man employed as an operator has also gradually changed:

1. He is, on average, better educated. Some organisations employ graduates as operators - although this is not the norm.

2. Young operators may never experience plant operation in the way that their older colleagues did prior to moving to jobs inside the control room. On some plants it is policy that operators spend a considerable period outside before taking up control room duties. Even when this

* Department of Chemical Engineering, Loughborough University of Technology

occurs the operators are unlikely to become really familiar with the plant because a small operating team will now run a large plant.

Fault diagnosis is often the most difficult task that the operator performs. His skill in performing this task is rarely monitored and he often receives no training to improve his diagnostic skills (but see references 1 to 6 for reports of progress in this area). The author knows of no long-term studies of process operator diagnostic performance so quantitative observations cannot be made. However it appears that:

1. There is no evidence that the operator's diagnostic skills have significantly improved or deteriorated. (We still know of some incidents attributable to diagnostic error but there is no sudden preponderance of them).

2. There is no evidence that changes in technology of plant instrumentation have made any significant impact on the support available for fault diagnosis.

The author stresses that these are only personal views and welcomes any evidence for or against.

It is quite clear from safety studies that on many plants there are event sequences which:

1. Have possible final outcomes which are either hazardous or very expensive.

2. Have a time scale such that timely intervention by the operator can prevent such undesirable sequences occurring.

There may also be automatic protective systems employed. The interaction between the various systems and the operator is shown in Figure 1.

The object of this paper is to examine the possible role of expert systems to improve the operator's diagnostic performance.

## EXPERT SYSTEMS

In the most general terms an Expert System is simply a computer program that is able to give advice on problems of a specific type. The problem area covered is often known as the "domain" of the system. The system formulates its advice on the basis of data for the domain provided by the "expert". The data is normally provided as a set of "rules" which can be modified or replaced as knowledge about the domain is accumulated. The Expert System is also normally capable of explaining its line of reasoning to the user in terms of the rules that it has used. The system works by asking the user a series of questions relevant to the problem. In a well-designed system the questions will rapidly become specific to that part of the domain in which the particular problem lies. (For instance a good Expert System for diagnosing failure to start a car engine would not start asking questions about the state of the battery after being told that a good spark was obtained at the plugs but it might very well ask about the battery if no spark was obtained at the plugs). These characteristics of Expert Systems are summarised in Table 1.

Many existing computer programs have some or all of the properties specified above and are hence in some sense "Expert Systems". It might therefore be considered that the terminology is rather superfluous. This is to some extent true and there is no doubt that much of the current popular interest in Expert Systems stems from a strong interest in their intended use as part of the Japanese 5th generation computer project - rather than because of an independent recognition of their value. This type of system does however have a well-defined and (in computing terms) long-established line of research effort by respected Artificial Intelligence researchers and is therefore not simply "trendy". Perhaps the least satisfactory aspect is the spurious credibility that might be implied by the coined title of "Expert System" - it may well be that a potentially useful tool could fall into disrepute by being associated with a small number of badly-designed systems or well-publicised failures. (Even the word "expert" is not universally acclaimed to be a complimentary term!)

Michie (7) contains more detailed information about the history of Expert Systems and their more well-known applications in medical diagnosis, geology etc.

No computer program can be a panacea for all of our problems and, in particular, 3 points should be borne in mind about Expert Systems:

1. They appear to be bestsuited to applications where many possibilities exist.

2. The result, when given by the system, is often not very surprising - but it may not have been easy to come to the same conclusion without the system (e.g. reason for car fault may seem obvious after mechanic discovers it).

3. The system is only as good as its data (the rules) - some experts are better than others!

## PLANT FAILURE DIAGNOSIS

Failure diagnosis is a skill that few people possess naturally (and those that can do it naturally are often not good at explaining how). Some or all of the following factors apply to most diagnostic problems:

1. The possible causes of failure may be many and varied.

2. The person called on to carry out the diagnosis may be unfamiliar or unable to recognise some failure mechanisms.

3. The operator is presented with the problem of selectively abstracting the relevant data from all the other indications. He does not know what is relevant until he has identified the fault - a "chicken and egg" problem. This is also strongly related to the post mortem comment:

   "......it was all there on the panel, if only the operator had seen it......"

4. The operator may be under considerable pressure to take the right action quickly - because some faults rapidly lead to hazardous and/or consequences.

5. The displayed symptoms may appear in a confusing order or may be misleading due to instrument faults (e.g. measuring device failed or pen-recorder stuck).

6. The operator may have had no training in fault diagnosis or the training may have been unsuited to the task.

7. The operator may be unfamiliar with the behaviour of the plant under the particular unsteady - state conditions prevailing.

It is not the intention of this paper to follow-up these points here - the list above is given to give some indication of the difficulty of the task.

On a properly-designed plant there are 3 basic levels of protection:

1. The control systems

2. The alarm systems

3. The trip systmes

Lawley's (8) approach to hazard analysis emphasises these 3 levels. A single plant variable may be part of all 3 levels of protection - although not all hazardous states fall into this system (e.g. a toxic gas concentration in a plant area accessible to plant personnel might only be monitored by alarms and/or cause a trip). Notice that the open-loop alarm system (requiring human intervention) comes <u>between</u> the automatic closed-loop systems. The Expert System fits nicely into the same place since it provides <u>advice</u> on which the operator can act. An Expert System which aims to diagnose Plant Failures from alarms and/or other indications is therefore a very natural addition to the current philosophy of plant operation and fault diagnosis.

## EXAMPLE APPLICATION

Consider the system shown in Figure 2. This simple plant section is intended to provide a downstream reactor with a constant flow of reactant. An early warning of flow failure is therefore a low level fault in the section shown. Table 2 shows some of the possible causes of TANK LEVEL LO (the causes shown are chosen to give a representative cross-section of fault types rather than being based on frequency of occurrence). Notice that Table 2 is essentially a list of plant (rather than instrument) faults. This distinction is deliberate - a fault occurring on the plant may appear quite differently to the operator if instrument faults are also present, because the instrument readings are his primary means for observations of the plant. Table 3 illustrates this by showing how one set of observations may be interpreted in terms of different plant faults when various instrument failures are assumed. Obviously it is unlikely that multiple independent instrument faults occur but even single instrument failures can cause confusion as shown in the table.

An Expert System can be easily set up to carry out the process shown in the example. The system has a defined set of plant faults which it attempts to match with the observed readings - assuming any instrument failures necessary to achieve the match. An experimental system at Loughborough was rapidly constructed to achieve this using the PROLOG language ,Clocksin and Mellish (9). The most likely fault can then be easily determined on a probabilistic basis. The user can always ask for the logic chain implied by a particular fault sequence in addition to the basic symptoms (see Table 4). He can also ask for any alternatives that are possible.

## DISCUSSION

The example given is only very simple and certainly does not conclusively prove that Expert Systems are a worthwhile diagnostic aid. However there are several points which do come out of the example which provide a basis for initial evaluation:

1. Some simple fault sequences are quite confusing when instrument faults are involved. The Expert System appears well-suited to unravelling these in a coherent manner.

2. The Expert System was easily set up even though the author was inexperienced with the programming language used (PROLOG). It would seem that an experienced programmer familiar with the language would be able to rapidly configure a good system - probably including better facilities and more efficient processing.

3. The Expert System only provides the same kind of aid as an Alarm Analysis or Disturbance Analysis System (DAS). However these systems are set up using conventional programming languages such as FORTRAN PASCAL, or RTL/2. The use of a non-procedural language has some advantages for this kind of application - because its facilities reflect the problem characteristics quite well - and should be investigated further. Some advocates of Expert Systems would see the "expert" as the operator whereas in DAS applications the data is provided by the design engineer. In the authors opinion it is not yet clear that the operator can provide better data than the designer - particularly for low-probability faults.

4. The characteristic Expert System features (data in form of rules, many possible alternatives, explanation of reasoning etc.) match the requirements of Fault Diagnosis very well. As Expert Systems evolve it may be possible to build a specific system rapidly from a general purpose "Expert System Shell". A "Shell" is simply an expert system with the domain-specific rules removed. It may then be used with the rules for a new domain. (Shells are available, e.g. EMYCIN, but their design does not really suit general-purpose use.)

5. It seems that the basic structure used is well-suited to the changing data inherent in a real plant situation (e.g. instruments on maintenance etc.), although special facilities for changing the rules in a secure way are needed. There is also an obvious requirement for rapid response in a process control environment. The limited trials so far carried out indicate that this is not a practical limitation.

6. Failure rate data can easily be built into the system if required (this has not been done at the time of writing).

There are also other considerations for any real application such as collection of the rule data, the "cut-off" point for low-frequency events, the size and organisation of data storage (this requires careful thought and is essential for good performance), the detailed design of displays and dialogues etc. These factors are very important but have not yet been considered in detail as part of the current project (Andow (10) and Andow and Rogenbauer(11) discuss some of these factors).

## CONCLUSIONS

The Expert System provides a possible framework for plant fault diagnosis. In particular it appears that the normal Expert System characteristics match the problem requirements well and hence provide a natural basis for a diagnostic facility. The simple experiments reported are not claimed to be conclusive but do show that a crude working system can be rapidly set up. The use of non-procedural programming languages (such as PROLOG) has been a major feature of the project and show that these languages are potentially very powerful - at least for this type of problem.

## ACKNOWLEDGEMENTS

The author wishes to acknowledge the support of the Science and Engineering Research Council (SERC) for the support provided for the related project "The Use of Expert Systems in Safety of Chemical Processes".

## REFERENCES

1. Rasmussen, J., and Lind, M., 1982, "A model of human decision-making in complex systems and its use for design of systems control strategies", Riso National Laboratory, Denmark, Report Riso M-2349.

2. Marshall, E.G. Scanlon, K.E., Shepherd, A., and Duncan, K.D., 1981, "Panel Diagnosis Training for Major-Hazard Continuous Process Installations". Chem. Engr.,.

3. Bainbridge, L., 1969, "The nature of the mental model in process control", in "Adaptive Man-Machine Systems. Display Design",

4. Lin, J.C., Guarro, S., and Okrent, D., 1982, "A deterministic methodology for application to a diagnosis oriented Disturbance Analysis System", Nuclear Engineering and Design, 72.

5. Rouse, W. B., 1978, "Human Problem Solving Performance in a Fault Diagnosis Task", IEEE Trans. on Systems Man and Cybernetics, SMC-8 (4).

6. Kragt, H., 1983, "Process Alarm Systems", Proc. 26th Annual Meeting of the Human Factors Society, Seattle, Washington, U.S.A.(1982).

7. Michie, D., (Ed), "Introductory readings on Expert Systems",

8. Lawley, H.G., 1980, "Safety Technology in the Chemical Industry: A Problem in Hazard Analysis with Solution", Reliability Engineering, 1 (2).

9. Clocksin, W.F., and Mellish, C.S., 1981, "Programming in Prolog", Springer-Verlag.

10. Andow, P.K., 1983, "Improvement of Process Alarm Systems", Eurochem '83, Birmingham, U.K.

11. Andow, P.K. and Roggenbauer, H., 1983, "Process Alarm Systems", Position Paper PP8305 of European Workshop on Industrial Computer Systems TC/6, (1983).

TABLE 1 - Expert System characteristics

| No. | Feature |
|---|---|
| 1 | Gives advice on problems within domain |
| 2 | Data provided by domain expert |
| 3 | Data in form of rules which can be modified |
| 4 | Capable of explaining own reasoning |
| 5 | Dialogue details conditioned by earlier responses |

TABLE 2 - Causes of TANK LEVEL LOW

| No. | Description |
|---|---|
| 1 | p1 LOW |
| 2 | f3 HIGH |
| 3 | CONTROL LOOP SETPOINT LOW |
| 4 | V1 OPEN AND V2 OPEN |
| 5 | p1 DISTURBED LOW AND CONTROLLER OFF-LINE |
| 6 | p1 DISTURBED LOW AND CONTROLLER STICKING |
| 7 | f3 DISTURBED HIGH AND CONTROLLER OFF-LINE |
| 8 | f3 DISTURBED HIGH AND CONTROLLER STICKING |

TABLE 3 - Alternative Interpretations of Indications

| Interpretation No. | | Level Variable (L) | Causes of Level Deviation | |
|---|---|---|---|---|
| | | | Flow (F1) | Controller State |
| 1 | Parameter Value | LOW | LOW | OFF-LINE OR STICKING |
| | Device State | NORMAL | NORMAL | |
| | Indication | LOW | LOW | |
| 2 | Parameter Value | UNKNOWN | - | - |
| | Device State | FAILED LOW | - | |
| | Indication | LOW | - | |

TABLE 4 - Explanation of logic used

<u>WHY IS LEVEL LOW?</u>

I DEDUCE THAT LEVEL IS LOW BECAUSE:

1. FLOW F1 IS LOW

AND

2. CONTROLLER IS OFF-LINE OR STICKING

<u>HOW?</u>

FLOW F1 IS LOW

BECAUSE

CONTROLLER OUTPUT IS LOW

AND

CONTROLLER OUTPUT IS LOW (AND SHOULD BE HIGH)

BECAUSE

CONTROLLER IS OFF-LINE OR STICKING

N.B. OPERATOR INPUT IS UNDERLINED

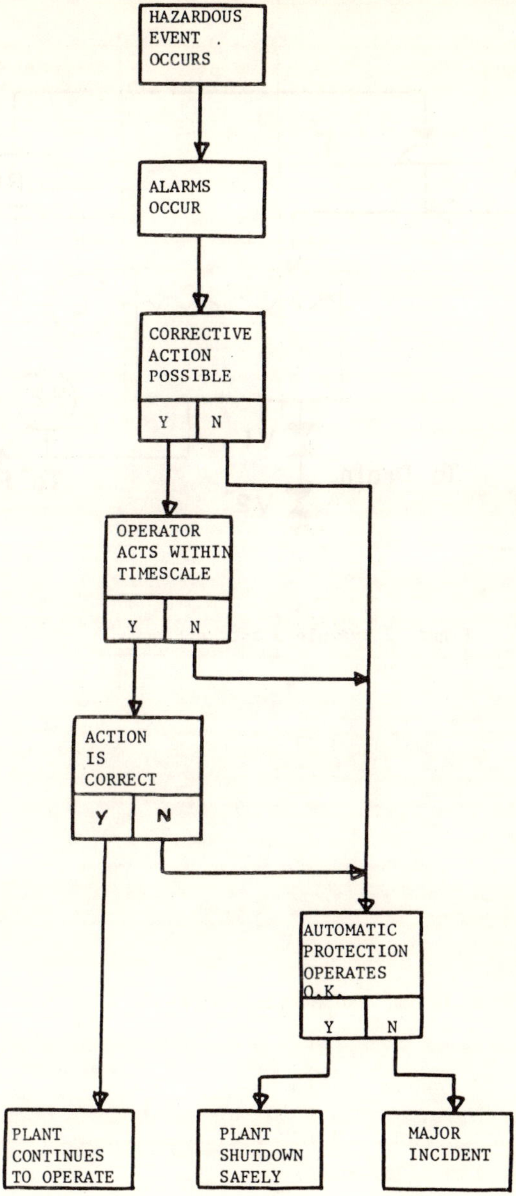


Figure 1 Simplified overview of alarms, operator actions and trips

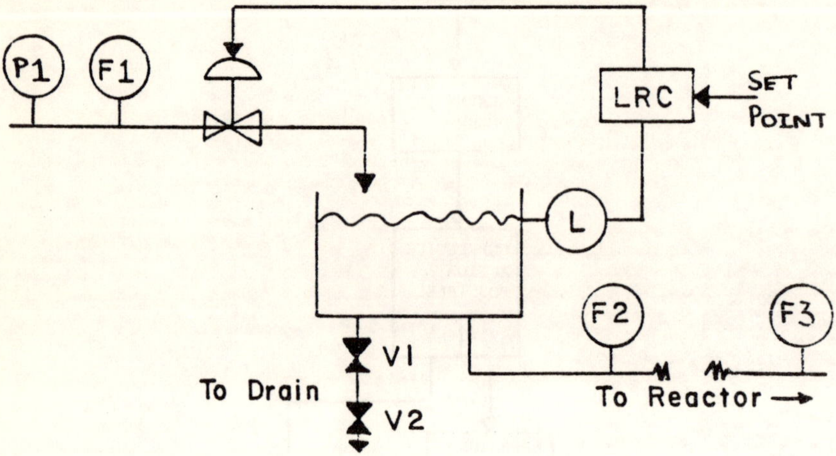


Figure 2 Example plant section

# AUTOMATIZED PROCESS CONTROL : THE ROLES OF COMPUTER-AVAILABLE INFORMATION AND FIELD-COLLECTED INFORMATION.

F. Daniellou (*), M. Boël (**)

In a computer-controlled oil-refinery, information on the course of the process is widely available from control room display screens. However, there is also a significant radio traffic between the control room and field operators.

An analysis of the matter of these calls makes it possible to highlight the role of information collected in the field in control room operators' decision-making.

On the one hand, some information on the current state of the process is not available from the computer. On the other hand, it appears that the control room operator frequently questions the reliability of computer-displayed data.

The elements presented in this paper on collective aspects of information gathering seek to make a contribution to future developments in the design of work organization, communication networks, and training programmes.

Many researches have been published during the last few years about information display in the control rooms of process industries. Most of them, however, isolate the problem of information display from the overall design of the control team's work organization.

The present paper, through the analysis of radio calls, emphasizes the collaboration between control room and field operators. It ends with some proposals concerning work organization and information display.

These results are part of a research carried out over one year in the control room of an oil refinery, which was automatized in 1978, and includes :

- a centralized automatic regulation of most valves,
- a computerized control aid system.

The unit in which the research was carried out is a refining section. It processes about 500 t of petrol per hour to make marketable products such as

(*) Laboratoire de Physiologie du Travail et d'Ergonomie du Conservatoire National des Arts et Métiers, 41 rue Gay-Lussac, 75005 PARIS -

(**) Groupe Communication et Travail, Université de Paris-Nord, Avenue J.B. Clément, F93430 VILLETANEUSE.

gasoline or gas. Subproducts are re-processed in other units. The number of regulated valves in the section considered amounts to about 600.

The team includes a board operator and a polyvalent operator working on several sections in the control room, 4 field operators and 2 foremen in the plant. The team is led by a chief-operator. The shift system is based on 5 shifts. However, the numbers are calculated over 6 shifts.

To understand the activity of operators controlling the process, professional conversations, radio calls and VDU consultations were recorded during 39 shifts, over a total period of 105 hours. The analysis of some records were submitted to operators in the course of interviews and their explanations noted down.

Taking in account the number of radio calls between to control room and field operators, we have attempted to analyse both the content of these calls, and their importance in the operators'activity.

It is extremely difficult to quantify the radio exchanges : it can be observed that in the course of some incidents, they are almost continuous, while in other periods only brief isolated calls are to be noted. Discussing the time distribution of both situations would need a precise presentation of the work load of the operators at any moment, which is not the main objective of this paper. Its aim is rather to give greater detail on the nature and the role of the radio calls.

## 1. INFORMATION REQUESTED BY THE CONTROL-ROOM OPERATOR

The control room operator frequently asks the field operators questions about indications that are either not available on the control room board, or which are not reliable enough.

### 1.1. Some Questions Deal With Information That Is Not Available From The Control Room Displays.

Many items of information relating to the state of the plant are not available from the board. Such is the case with :

. The state of devices not fitted with sensors. For instance : bypasses or safety valves. Most devices (pumps, automatic valves, exchangers) can by bypassed for maintenance needs. Apart from maintenance interventions, bypasses and safety valves can be temporarily used to locally modify the status of the devices. But relevant information can only be obtained through the intermediary of a field operator.

. Devices with only local warning lights. Thus the on/off status of a pump, its being driven by an electric motor or a turbine, single or dual operation, can only be cleared up on the spot.

. The progress of works being carried out on the plant. When taking over, the control-room operator is theoretically informed verbally and by means of an order-book of works in progress. However, he needs updated information, particularly about :

- the completion status of sensor-calibrations, so as to be in a position to evaluate the reliability of displayed information
- whether maintenance operators are present or not close to devices he is operating on.

1.2. Some Items of Information Are Not Available Soon Enough From the Board

This is especially the case with results produced by automatic analysers (chromatographs or distillators) after a time which can amount to ten minutes or more.

1.3. The Reliability of some Indicators is Questionned by the Operator

In the course of recordings, it is observed that the operator is frequently led to question certain values displayed : measures, valve-aperture, analyses.. As a matter of fact, sensors can break down, valves can seize up.
The observed frequence of malfunctionings does not amount to a high equipment failure rate. However, due to the number of regulation loops (about 600), the operator has to face failures of devices during almost every shift.

Thus, detecting faulty devices is not a peripheral activity, but rather is a central element of the operator's activity.

When the comparison of several displayed indicators leads him to doubts about the reliability of a sensor or a valve, he asks for a field check.

In all cases, information gained from the field is of a different nature than that displayed. This point will now be clarified.

## 2. INDICES FROM OUTSIDE

Field operators have at their disposal certain indicators of physico-chemical values, which are redundant with those displayed in the control room (e.g. tank levels, manometers). Comparing these two indicators makes it possible to diagnose a failure in a sensor.
However most indices gained on the spot are of different nature.

2.1. Some Field Indices have a Higher Reliability :

Thus seeing the aperture of a valve right on the spot makes it possible to remove the doubts about the value displayed.

2.2. Some Indices Are More Synthetic :

When adjusting a furnace or a boiler, the control room operator has at his disposal :

- the feed flow,
- the oxygen percentage in the furnace,
- the pressure value,
- the fuel flows,
- the material temperature,
- the pipe temperature.

The field operator can observe :

- the appearance of the flame,
- the features of the smoke,
- the noise made by the furnace.

These indicators enable him to immediately evaluate the result of an adjustment in the air flow, whatever the failures in the telemeasurement equipment might be, knowing that they are especially likely to occur in these very harsh conditions.

But, on the other hand, the field operator cannot act from outside on all parameter adjustments. These are the result of a long radio exchange between the control room and the field operators.

2.3. Somes Indices Are More Immediate

When the control room operator finds it too long to wait for the result of an automatic analysis, the field operator is requested to take a sample. The mere description of its colour is often enough to enable further adjustments in the refining column to be made.

2.4. Information Collected In The Field Permits The Continuity Of The Process to be Restored

For instance, if a flow displayed in the control room seems incompatible with the valve-aperture, the field operator can :

- check the valve,
- listen to learn whether the material is flowing past or not (the flow gauge may be out of order)
- follow the pipes to make sure that no hand valve has remained closed, whether pumps are running, etc ..

In face of the discontinuous nature of parameters delivered by sensors, the physical presence of an operator in the field makes it possible to restore the continuity of the process.

The analysis of indices collected straight from the devices highlights the importance of direct physical contact, as opposed to the remoteness and mediation which are the main features of a centralized control room. The impact of these exchanges on the operators' activity will now be discussed.

## 3. THE IMPACT OF RADIO EXCHANGES ON THE OPERATORS' ACTIVITY

The analysis of radio calls gives prominence to the fact that the operator does not only CONSULT information. Through a series of activities, which include VDU consultation, questions asked, interventions on the process and the analysis of their effects, he progressively CONSTRUCTS the information about the state of the plant. In some situations, he may need to collate several indicators of different types, and associate them with their reliability level (that is : what is the chance for this value to be faulty ?), and their age.

The elements supplied by field operators hold a central place in this progressive construction of information, since they make it possible for a decision to be taken among several interpretations of a disturbance.

For instance, when the control room operator notices that the temperature of a material cooled by a fan does not decrease sufficiently, he does not know whether the temperature-gauge is out of order, or whether the fan is malfunctioning. He sends the field operator to see if the fan is working properly.

. Collecting information in the field involves specific difficulties, moving around the plant, climbing up ladders, passing through hot areas, wearing gas masks in certain cases , which imply time.

The control room operator is able to foresee and take into account the time required, which means that he knows the geographical layout of the plant in a very detailed manner.

For example, when there is a query about a tank level, the information memorized and verbalized is not "the last time I've consulted the display, the level was 80%", but rather something like : "Ten minutes ago, the displayed level was 80%, which the field operator has confirmed. The valve was opened, the bypass too, but the level did not decrease. So I've sent the operator to follow the line downstream and see whether a hand valve might be closed. It still does not decrease. Ten minutes are time enough to do that, I've got no answer, I will call him back".

. The exchanges between the control room and the field operators cannot be understood merely as orders given by the control room and carried out in the field. It is much more a matter of collaboration in the construction of information, each having at his disposal unique indices. Any operator can be at the initiative of detecting an incident.

. It can be observed that the control room operator does not only transmit immediate information about current actions, but also expresses indications about successive stages of his control activity, such as "we have just switched over the feed". It is probable that the watch kept on the devices by field operators will be reorientated by this warning, and some parts of the plant monitored in a closer manner, without the foremen needing to point it out explicitly. It can be assumed that the field operators update their representation of the overall process through listening to all radio calls, including those which are not meant for them. This has already been highlighted concerning tug captains in a landing operation (BROUARD & al., 1983).

It appears that radio exchanges between the control room and field operators may not be confined to calls isolated from each other. On the contrary, calls come within the scope of the different operators who are following the same problems. Each operator has a different role, different activities, different constraints, and only part of the information.

However the successive cycles of information collecting and actions, which permit the solution of an incident bring into play the convergent activities of all members of the control-team.

## 4. CONCLUSIONS

Some points highlighted by the analysis of radio calls have to be taken into account when designing a new control room or work organization.

4.1. One of the central components of the operators' activity is to detect and diagnose failures in the automatic regulation, through a progressive construction of information, including different indices and actions on the process.

Direct collection of information on the spot, and the control room operator's knowledge of the lay-out of the plant, enable an early detection of malfunctionings.

4.2. This raises the question of the organization of work-sharing between operators. A work organization which rigidly imposes the permanent presence of control room operators at the board does not allow them to update their concrete knowledge of the installation. It may lead to a more difficult diagnosis of failing sensors or devices.

Maintaining a concrete knowledge of the plant is one of the essential stakes in work organization design.

4.3. The complementarity of operators' activities is made easier by everybody's knowledge of the difficulties encountered by his colleagues. This raises the question of team training, as an addition to the isolated individual training received by operators in the course of their career. The team mentioned does not only include the production team, but also the maintenance operators (electricians, calibrators, etc.) who daily work together with the former.

4.4. Whatever the attempts made by designers to improve the reliability of sensors and devices, it may not be assumed to reach 100%. This assumption is often implicitly made. On the contrary, an investigation of the detection and diagnosis made by the operators may lead to diagnostic aids.

4.5. In particular, owing to the convergence of control room and field operators' activity when detecting failures, it is probable that centralizing all displays only in the control room is not appropriate. Restoring some indications in the plant itself, in carefully cho sen place and number, would make for a faster detection of contradictory values.

4.6. Obviously, the technical quality of transmissions has to be closely investigated in all circumstances. It has been observed that hail storms led both to disturbances in many regulations AND to a radio black-out. This gives prominence to the fact that it is necessary to design reserve networks, from a precise analysis of where control room has to be called for in emergency.

4.7. To sum up, the current trend towards increasingly laying emphasis on human factors in process control should not be confined to control room design. It must extend to the overall design of the work organization of the whole team, whose collaboration is a condition for the smooth running of the system, both from the economic and the working conditions points of view.

## REFERENCE

BROUARD, J., DEUTSCH, C., ROUTIN, M., 1983
Le rôle des communications verbales radio dans la coordination entre les différents acteurs d'un système "hommes-machine" : l'accostage des grands navires.
Colloque "Techniques Nouvelles et Ergonomie", Valenciennes (France), 31 Mai-2 Juin 1983.

# APPLICATION OF HUMAN RELIABILITY ASSESSMENT TECHNIQUES TO PROCESS PLANT DESIGN

D.E. Embrey*

The application of human reliability assessment techniques to process plant design is first considered from the point of view of qualitative modelling, to identify the human errors which would affect safety, production or availability, and the design changes which could reduce these errors. In order to allow cost benefit analyses, quantification of the likelihood of error is necessary. Techniques are described which allow the numerical effects of design changes on human reliability to be assessed.

## INTRODUCTION

The objectives of process plant design are to produce a system which maximises production and safety goals at minimum cost. The attainment of these goals will be strongly influenced by the extent to which operability and maintainability are built into the design such that the likelihood of human errors in these areas are minimised. Human reliability assessment (HRA) techniques can be applied during the design stage and during plant modifications and retro-fits as an aid to achieving these objectives. One of the advantages of the HRA approach is that it provides the designer with techniques which will indicate the relative cost effectiveness of different solutions to the objective of maximising human reliability and thereby minimising expensive errors. This is a particularly useful feature in current design environments where expenditure has to be explicitly justified on a cost-effectiveness basis. Another advantage of the HRA approach is that it enables the relative impact of hardware aspects of the design to be compared with other areas of the total man-machine system design, e.g. improved procedures, training etc. to establish the least expensive route to maximising human reliability. Finally, the qualitative aspects of HRA, the identification and modelling of potential human errors, can be a very effective means of raising the consciousness of the design team with regard to identifying those aspects of the system which are likely to induce errors, and which may be amenable to design solutions.

* Human Reliability Associates Ltd., 1 School House, Higher Lane, Dalton, Parbold, Lancashire WN8 7RP, ENGLAND.

## MODELLING POTENTIAL HUMAN ERRORS

The first stage of the assessment process involves describing the role of the operator in situations where his actions (or errors) have an impact on the safety, availability or production aspects of the plant. Typical areas of interest are control room based activities, such as normal production operations, and the management of accident situations, together with activities which may occur remote from the control room, such as maintenance (particularly of safety systems).

The process of identifying and modelling potential operator errors is best carried out as a joint process between the hardware systems designer and the human reliability specialist. This is so that the potential impact of both errors of commission and omission on the system as a whole can be evaluated. In addition, the plant designer and HRA specialist will be able to specify in detail the role of the operator and the nature of the process with which he will have to interact. The term modelling is used in a similar manner to its application in reliability assessment, and refers to the analysis of the various ways in which a particular human error could occur.

A number of tools are available to assist in the identification of potential failures. The approach which is used will depend on the type of situation being assessed. In some cases, the operator's role is rigidly proceduralised, for example in routine maintenance operations where discrete steps are specified by means of a checklist. Procedures may also be the basis for operator actions in routine process plant control, or for the management of clearly defined accident situations.

However, the growing application of computers for routine control functions means that apart from the maintenance and testing area, the control room operator is increasingly taking on the role of a monitor and system manager. His function, therefore, is primarily to utilise higher level skills such as decision making, diagnosis, trouble shooting and planning. These skills are also important in the management of abnormal, low probability events which cannot easily be anticipated, but which have a very high hazard potential. Different analytical approaches are necessary for proceduralised actions and for situations where decision making functions predominate.

### Human Reliability Modelling in Proceduralised Situations

The area of proceduralised behaviour, referred to as 'rule-based' behaviour in the classification developed by Rasmussen (1) (see following section) can be analysed using a simple task analysis approach such as that shown in figure 1.

The task step which is inserted in column 1 is a discrete element of the overall task which is assigned to the operator. Frequently, the entry in the task step column corresponds to a specific step in the procedures. The 'inputs' column describes the initiating cue for the action, which may be a control room alarm or indicator, or an item on a checklist, in the case of a maintenance procedure. The 'outputs' column describes the observable actions of the operator, and the controls which may be involved in these actions. The feedback that signifies the completion of the task step is identified in the fourth column. Feedback information, since it is an input to the operator, could be included in column 2. However, since feedback (or its absence) is a particularly important aspect of error reduction and causation it is useful to analyse it separately. Column 5 describes potential errors which could occur in the task step being considered, and the following column considers the safety or availability implications of the error (information from the hardware engineer will be important here). Information concerning typical error patterns gained from the analysis of event reports can be factored into the error potential column. For example, the omission or reversal of procedural steps is most likely to occur when they are isolated from the main sequence of the procedure in the sense of not being obviously related to preceding or following steps, e.g. forgetting to return a valve or switch from the 'test' to the 'operate' position.

A very high proportion of errors that occur are recovered, providing the system is 'forgiving' and the operator is provided with the adequate feedback from the process. Column 7 therefore describes the cues which indicate that a particular error has occurred, and possible actions that the operator can take to recover the situation.

The final column considers the design implications of the earlier columns. These may include the use of hardware interlocks if the consequences of the errors are particularly serious, changes in the man-machine interface (e.g. moving or relabelling ambiguous valves, making error recovery cues more obvious) or changes in procedures or training design.

Figure 1 constitutes a simplified example of the type of analysis that is possible, in a chemical plant context. The task step which is analysed would be one of several which would form part of a particular task assigned to the operator. Considerably more detail could be incorporated in the table, and other failure modes could be identified, e.g. the operator could fail to observe the temperature reading for pan Q on the chart recorder. A very similar analytical format could be used for other sequential proceduralised tasks such as those found in maintenance and testing.

Where the various task elements form an ordered sequential sequence, moving forward in time, the effects of errors at each step can be modelled using an event tree format, McCormick (2) (figure 2). Event trees have primarily been developed in a hardware context to model the possible failures of various systems that should operate sequentially following an initiating event in an accident sequence. However, they can equally well be applied to modelling human reliability in sequential, proceduralised situations where discrete actions can be readily identified. Event trees are of particular importance in the quantification of human reliability, as will be discussed in subsequent sections. The event tree gives a global view of the various failures which could occur in a particular task. For design purposes a more fine-grained analysis will often be necessary, in which a further decomposition of the causes of failure are made. One way of representing this is to use a fault tree which shows how different sources of failure can be combined together using 'and' gates and 'or' gates, to give rise to the overall failure in the sequential step being considered. Much of the information contained in a fault tree can be described verbally using the format shown in figure 1. For example, an additional failure mode of the step being considered in figure 1 would be failure to detect the indication on the chart recorder that signalled the need for the reactant to be added. This could be added to the information already contained in column 5 of figure 1, with appropriate additional entries in columns 6 to 8. For simple sequential tasks it is unlikely that a fault tree with more than one level of decomposition would be necessary.

The tabular format of figure 1 is unsuitable for describing complex situations, e.g. accidents where the operator does not interact with the system in a linear, sequential manner. Also the format of figure 1 does not contain any reference to higher level cognitive processes e.g. thinking, decision making, and planning. For more complex situations other approaches are required which consider these processes explicitly.

## Human Reliability Modelling in Complex Situations

In complex situations there may be many interrelated parallel activities involved, especially in accident management situations, and cognitive functions such as decision making become particularly important.

Because of the impossibility of modelling the situation in the detailed manner which is feasible for proceduralised situations, it is useful to represent the functions that the operator has to perform at a higher level than in the simple case. As an aid to facilitating communication between the designer and the HRA specialist, the Operator Action Event Tree, (OAET), Brown et al. (4) can be used, (figure 3). This indicates the functions that the operating team is expected to perform in relation to critical states of the system. The technique is utilised in severe accident sequence analyses in nuclear power plants, but can also be applied in other systems. At each of the critical states the following information is derived:

a) Postulated sequence of events : This gives a brief description of the failure events which produced the state and their implications for the failure of critical functions, e.g. cooling.

b) Required operator responses : This is a brief description of the appropriate operator response to the state, and includes confirmation, diagnostic and action items. A distinction is made between pre-planned (proceduralised) actions, essential actions, and those which can be delayed.

c) Key symptoms : The major symptoms which are associated with the state are identified and described. Changes in parameters associated with the current state are delineated.

d) Uncertainties : Aspects of the required operator responses or the state symptoms which are uncertain (because of the complexities of the situation) are described here.

e) Potential substates : Because of knowledge gained in steps (b), (c) and (d), the structure of the original OAET may be revised at this stage to include additional states or substates.

f) Additional comments : Any significant comment that does not fit into the above categories is included here.

The OAET defines the operator actions associated with the critical states of a sequence, and provides useful data for evaluating the likelihood of current diagnosis, and the control actions that the operator will have to make in relation to these states. It does not, however, indicate how the operators might fail, or for that matter how operator errors unrelated to these states could give rise to other critical situations. The chief value of the OAET is as an adjunct to a hardware orientated safety or hazard analysis where failures of various critical systems are postulated, and the operators' response to these failures needs to be evaluated.

Another way to analyse human reliability in complex, unstructured situations is to work from the critical decisions/ actions that the operator will be required to make. This orientation differs from that of the OAET, in that it is operator-centred rather than function-centred. A critical decision/action (CDA) is defined in terms of its consequences. If a CDA fails it will have a significant effect on safety, production or availability. In the case of the routine sequential actions considered in the last section, the only critical decision made by the operator is usually whether or not a particular procedure is applicable to the current situation. Once a choice has been made, the operator follows the decisions made by the originators of the procedures.

A format has been developed by Human Reliability Associates to assist in the analysis of CDAs in complex situations. The first stage of this procedure consists of identifying the CDAs in the context of significant changes of state in the system being analysed. The approach differs from the OAET in that it does not simply consider the required actions in response to critical system states, but is concerned with the decision making which underlies these actions. Having identified the CDAs that are likely to be associated with the situation being analysed, each CDA is then considered from the point of view of the decision/ action elements that it contains, and the potential failures that can occur at each of these elements.

The decision/action elements that are considered are based

on a model developed by Rasmussen (1). In simplified form, this can be represented as shown in figure 4.

This model is applied in the following way. The activation element corresponds to the initial alerting of the operator that action is required, e.g. by the onset of a pattern of symptoms or alarms. In most cases, with an experienced operator, these will be recognised as a familiar pattern, for which a simple pre-programmed action is available. Hence, the operator will normally execute the action appropriate to the symptoms, as shown by the arrow in the lower segment of figure 4. Behaviour of this type is called skill based behaviour, because it is characteristic of skilled, experienced operators. The capability to work in a skill based mode comes from experience with a specific plant, where particular configurations of frequently occurring symptoms become associated with particular causes. Of course, this can also be the origin of errors if the operator produces a stereotyped response to a pattern of indications which is superficially similar to the familiar pattern but which actually arises from a different cause.

If, as in some cases, the operator (or operating team) does not immediately recognise a pattern of indications, this will normally initiate a process of scanning and information collection from various sources (including procedures, if these are symptom-based). This will then usually lead to the selection of an appropriate procedure (either formalised as an explicit plant procedure, or from the operator's memory) which will be executed to achieve the desired result. This mode of behaviour is referred to as rule-based, since it involves explicit rule-following. Rule based behaviour will normally occur somewhat less frequently than skill-based behaviour, which is typified by 'man-in-the-loop' process control where minor adjustments may be made frequently to maintain the system within required limits.

If, even after the information gathering carried out during the integration phase, a diagnosis cannot be made which leads to the implementation of a specific procedure, the operators enter the realm of knowledge based behaviour. In this mode, the decision maker cannot rely on past experience, but has to formulate new hypotheses from scratch, probably based on general engineering principles involving abstract concepts such as mass and energy flows, heat sinks and sources, and transfers of inventory. Once an interpretation has been arrived at, and a suitable strategy has been formulated to achieve a desired goal (e.g. a safe stable state) an appropriate procedure has to be selected to reach this goal. This may involve the use of existing procedures or the improvisation of new ones. Finally, the execution of the procedures will involve concrete actions such as opening and closing valves, turning on pumps etc.

It should be emphasised that in practice, the types of behaviour described above occur as a series of tightly coupled, iterative steps. An important aspect of the model is that it is hierarchical. Thus, knowledge based behaviour has to proceed to rule based and then skill based behaviour, to actually execute the chosen procedure.

From the analysis of event reports and other sources, characteristic types of error can be associated with the different decision elements which may occur within each CDA. Figure 5 gives a more detailed breakdown of these elements and their associated errors. This approach can be used by the analyst to help postulate likely error modes. It is possible to provide more detail about the causes of errors at each of the decision elements described in figure 5, in a form which can be readily utilised by designers. Pew et al.(4) have developed a series of 'Murphy Diagrams' (named after the well known Murphy's Law : "If something can go wrong, it will"). Each decision element in the Rasmussen model has an associated Murphy diagram, an example of which is shown in figure 6 for the decision element 'Task Definition and Selection of Strategy (Goal State)'. It can be seen that this particular Murphy Diagram suggests two 'proximal' i.e. primary causes of the failure 'inappropriate strategy selected'. Each of these causes are then considered in terms of other causes which could influence or give rise to the primary cause. The Murphy Diagram can be of considerable value to the designer because it suggests specific causes of errors which will be amenable to design solutions.

It may seem a formidable task to apply the approaches discussed in this section to every CDA that is likely to occur in a complex event. However, only a relatively small number of decision elements will be associated with each CDA in most cases, which means that the process of analysis is reasonably manageable.

### Discussion

The techniques outlined in the preceding sections demonstrate that human reliability orientated analyses of both proceduralised and complex situations produce a number of insights which can be directly translated into design recommendations for error reduction. These approaches can be applied at both the early design stage, and also when it is necessary to analyse an existing system for the purpose of error reduction. The techniques also form a useful complement to more hardware orientated approaches such as fault and event tree techniques, failure mode and effects analysis, and hazard and operability studies.

## HUMAN RELIABILITY QUANTIFICATION AND DESIGN

The identification of the nature of the potential human errors in a situation, together with possible design solutions for reducing the likelihood of these errors, is an important aspect of HRA and process plant design. However, as discussed in the introduction, in the present economic climate there are usually severe cost constraints which mean that any design proposals have to be justified in cost effectiveness terms. In the human reliability context it is necessary to demonstrate that any proposed expenditure to reduce human error will produce substantial benefits in terms of reduced losses, increased availability or achieved by quantifying the reduction in human error which will be achieved by a given expenditure on design modifications.

## THERP - A Decomposition Approach to Quantification

A number of approaches are available for the quantification of human reliability, (see for example, Embrey (5)). The best known methodology is the Technique for Human Error Rate Prediction (THERP), which is described in detail in the 'Handbook of Human Reliability Analysis', Swain and Guttman (6). In THERP, a task for which quantification is required is broken down into a series of sequential task elements such as opening valves, reading dials, or consulting checklists. This is used to produce an event tree which models the likely failures which could occur. Swain and Guttman (6) contains tables of error probabilities for task elements which are typically encountered in nuclear power plants and other control room situations. These elemental failure probabilities are largely derived from the individual judgements of the compilers of the handbook. The task element probabilities are modified to reflect the situation specific characteristics of the task to which they are being applied and are then inserted into the event tree. The modified task element probabilities are then combined to give the overall probability of failure for the task.

For the purpose of producing design recommendations, it is possible to examine the THERP event tree to identify the particular elements which are major contributors to the overall probability of failure. These aspects of the task would obviously be the areas to which design changes would be directed. However, the THERP approach does not allow true sensitivity analysis to be carried out. It is not possible to determine, either at a task element or an overall task level, the extent to which the probability of error would be changed given, for example, changes in the interface as compared with improvements in training or procedures. Another disadvantage of THERP is that it is primarily applicable to proceduralised rule-based sequential tasks. It does not easily address the complex knowledge-based situations which were discussed in an earlier section where decision making and problem solving are important aspects of the human's role.

## Wholistic Techniques for Quantification

From the point of view of their application to design, there are other techniques available for human reliability quantification which overcome some of the disadvantages of THERP. The main characteristic of these approaches is that the task of interest is not decomposed into a series of sequential elements to which data are applied from a data bank. Instead, the task is evaluated as a whole and quantified via the structured application of explicit numerical judgements from groups of experts with plant experience and other relevant knowledge.

The most developed of these approaches is the Success Likelihood Index Methodology (SLIM) which is described in Embrey (7). Recently this technique has been implemented using an interactive computer program called MAUD (Multi-Attribute Utility Decomposition), the resulting technique being known as SLIM-MAUD, Embrey (8). In SLIM-MAUD, a set of up to eight tasks for which quantification is required are assessed as a group. The interactive computer program elicits from the judges the parameters known as Performance Shaping Factors (PSFs), which are felt to be the major determinants of human reliability for the task set being evaluated. Typical PSFs include training, procedures, quality of the interface design, etc. Having elicited the critical PSFs, MAUD then asks the judges to numerically rate the quality of these PSFs for each of the tasks being assessed. Quality, in this context, is the extent to which the various PSFs are either good or bad in terms of their effects on reliability in each of the tasks being evaluated. Finally, MAUD obtains the relative importance or weight that the judges assign to each of the PSFs in terms of their overall effect on human reliability. Thus, the ratings are different for each task, but the relative weights apply to all the tasks in the set being assessed. MAUD then sums the product of the weights and the ratings for each task, thereby producing a numerical index in each case called the Success Likelihood Index (SLI). The SLIs represent the position of each task on a scale of likelihood of success. The SLIs can be converted to probabilities of success (or failure) for each task if at least two tasks with known failure probability are included in the set being evaluated. This enables the SLI scale to be calibrated to form a probability scale.

From the point of view of providing design recommendations, the numerical weights generated by SLIM-MAUD enable the relative effects of different design parameters (if these are explicitly included in the PSF set) to be determined. Thus, it is possible to identify which design factor has the greatest effect on the overall probability of success for a particular set of tasks. This enables resources to be deployed most effectively in making design changes which will have the maximum impact on human reliability. The technique also enables predictions to be made regarding the actual changes in success probability that will result from specific improvements in PSFs such as training or the design of the interface. Another major advantage of the SLIM-MAUD approach is that it can be applied to complex situations involving decision making and problem solving, as well as to proceduralised situations.

The other major wholistic technique is the Influence Diagram. This is a method for the quantification of human reliability which involves modelling the network of influences which combine at various levels to determine the expected probability of error for the task being assessed. The Influence Diagram is described in detail in the human reliability workshop proceedings (this conference) and in Embrey (9). One particular advantage of the Influence Diagram is that it allows complex interrelations between influences to be evaluated. For example, it is possible to determine the effect on human reliability of a situation where the procedures, training and interface design are all good compared with the same situation where training is poor, or where both procedures and training are poor but the interface design is good. The Influence Diagram requires more time and resources to exercise compared with SLIM-MAUD, but it has a correspondingly greater value as a means of raising the consciousness of the design team with regard to the interrelationships between different parameters in terms of their impact on human reliability. Both SLIM-MAUD and the Influence Diagram have a similar sensitivity analysis capability.

## CONCLUSIONS

The object of this paper has been to demonstrate that human reliability assessment approaches are not only applicable to the relatively specialised area of probabilistic risk assessment, but can also be a valuable aid to the designer in identifying potential human errors and indicating the most cost-effective strategy for reducing these errors. It is hoped that the application of human reliability assessment techniques to providing numerical estimates of the benefits which can accrue from the use of ergonomics in process plant design will enable designers to produce safer and more productive systems.

## References


1. Rasmussen, J., 1980, "The Human as a Systems Component". In : Smith, H.T., Green,T.R.G. (eds) Human Interaction with Computers, (London : Academic Press).

2. McCormick, N.J., 1981, Reliability and Risk Analysis, (New York, Academic Press).

3. Brown, R.G., Von Herrman, J.L., Quilliam, J.F. 1982, "Operator Action Event Trees for the Zion 1 Pressurized Water Reactor". NUREG/CR-2888, EG&G Idaho Inc., Idaho Falls, Idaho 83415, U.S.A.

4. Pew, R.W., Miller, D.C. and Feerher, C.E., 1981, "Evaluation of Proposed Control Room Improvements through Analysis of Critical Operator Decisions". Report No. NP-1982, Electric Power Research Institute, Palo Alto, California, U.S.A.

5. Embrey, D.E., 1976, "Human Reliability in Complex Systems : An Overview". Report R10, National Centre of Systems Reliability, United Kingdom Atomic Energy Authority, Wigshaw Lane, Culcheth, Warrington WA3 4NE, England.

6. Swain, A.D., and Guttmann, H.E., 1983, "Handbook of Human Reliability Analysis Applications". Report NUREG/CR-1278, Sandia National Laboratories, Albuquerque, N.M. 8715, U.S.A.

7. Embrey, E.D., 1983, "The Use of Performance Shaping Factors and Quantified Expert Judgement in the Evaluation of Human Reliability : An Initial Appraisal". Report NUREG/CR-2986, Department of Nuclear Energy, Brookhaven National Laboratory, Upton, N.Y. 11973, U.S.A.

8. Embrey, D.E., Humphreys, P.C., Rosa, E.A., Kirwan, B., and Rea, K., 1984, "SLIM-MAUD : An Approach to Assessing Human Error Probabilities Using Structured Expert Judgement. Volume I : Overview of SLIM-MAUD". Report NUREG/CR-3518, Department of Nuclear Energy, Brookhaven National Laboratory, Upton N.Y. 11973, U.S.A.

9. Embrey, D.E., 1983, "Modelling and Quantifying Human Reliability in Abnormal Conditions". Proceedings of the 4th National Reliability Conference, Volume I. National Centre of Systems Reliability, Wigshaw Lane, Culcheth, Warrington WA3 4NE, England.

| 1<br>TASK STEP | 2<br>INPUTS<br>(cues for action) | 3<br>OUTPUTS<br>(action) | 4<br>FEEDBACK | 5<br>ERROR<br>POTENTIAL | 6<br>SYSTEM<br>IMPLICATIONS | 7<br>ERROR<br>RECOVERY | 8<br>DESIGN<br>IMPLICATIONS |
|---|---|---|---|---|---|---|---|
| Addition of reactant B to Pan Q from charging vessel 4. | Temperature in Pan Q greater than 200°C (chart recorder 6, trace 2). | Open valve 12B. | Weight reading from charge vessel 4 indicates transfer complete. Temperature in Pan Q increases to over 250°C in 10 minutes. | Could mistake valve 12A for 12B. (Close together on panel). | Hazard: 12A will cause reactant to be added to Pan Q causing dangerous exotherm. Temperature will rise to 450°C in 5 minutes. | If 12A operated by mistake, weight reading from charge vessel 4 will not change. Rapid temperature rise in Pan Q. Operator can close valve 1B to prevent transfer if mistake detected within 2 minutes. | Consider interlock. Clearly separate and re-number valves 12A and 12B. Insert procedural check for operator to monitor weight change in charge vessel and temperature rise in Pan Q. |

Figure 1 : Typical Entries in Task Analysis for Human Reliability Assessment in a Proceduralised Situation.

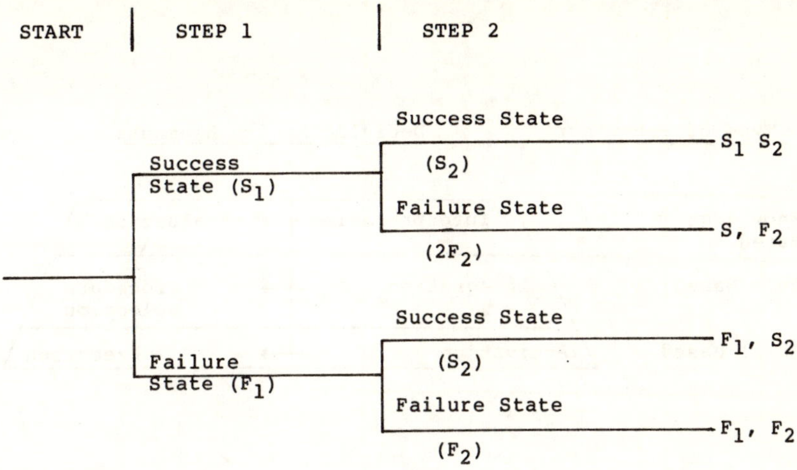


Figure 2. Event Tree for Proceduralised Situations

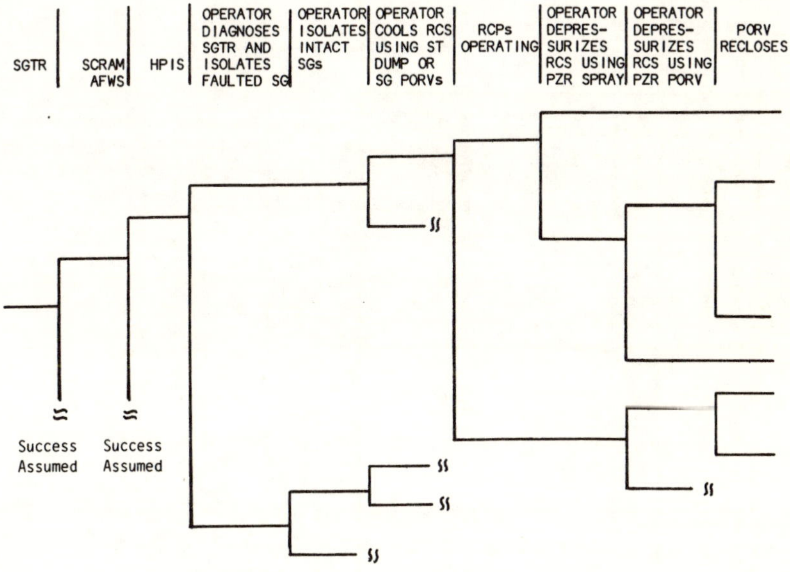


Figure 3. A Typical Operator Action Event Tree - after Brown et al. (4)

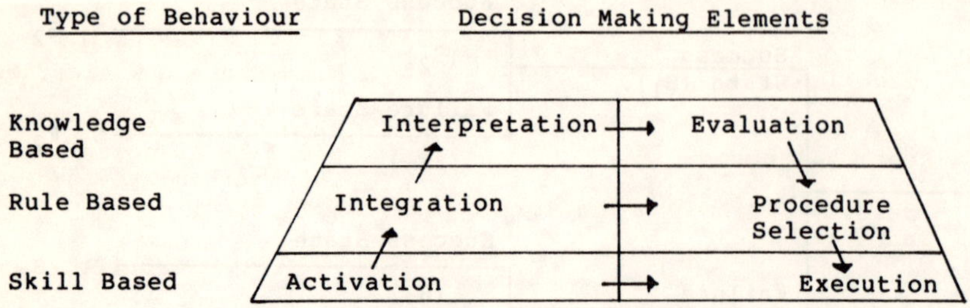


Figure 4. Decision Making Model from Rasmussen (1)

| DECISION/ACTION (Element) | NOTES | TYPICAL ERROR PATTERNS |
|---|---|---|
| a) Action/ Detection | Initial Alert | Distraction/Absent-Mindedness/Low Alertness |
| b) Observation/ Data Collection | Observation | Unjustified Assumptions/ Familiar Associations/ Information Overload - |
| c) Identify System State | What is Wrong? | Time Delay |
| d) Interpret Situation | What is the Cause? What if no Intervention? | Failure to Consider Side Effects/ |
| e) Evaluation of Alternatives | What to do? What are the Consequences of Alternatives? | Focussing on Main Event |
| f) Task Definition/ Selection of Goal State | What State Do We Want To Get The System To? | Wrong Task May be Selected due to Shortcuts in Reasoning and Stereotyped Response to Familiar State |
| g) Procedure Selection/ Formulation | How do we Get There? | Procedural Steps Omitted/Reversed (Particularly if 'Isolated') |
| h) Procedure Execution | Carry out Actions | Reversals of Direction or Sign (Up/Down, Left/ Right) Habit Intrusion |

Figure 5. Procedures to Model Human Error Modes

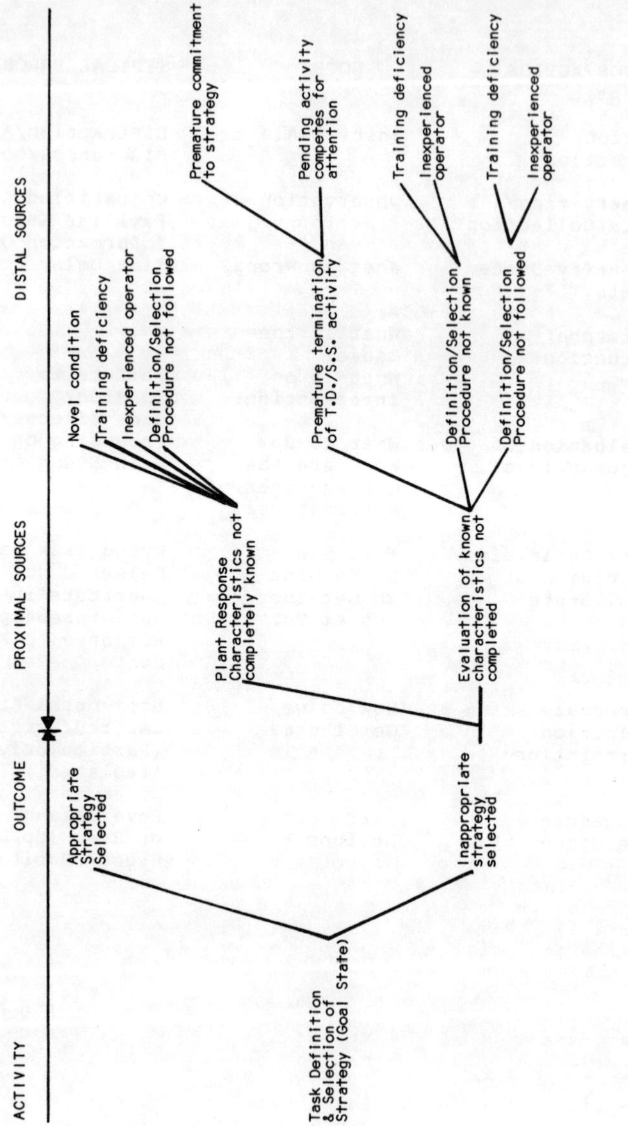


Figure 6.  Murphy Diagram for Task Definition and Strategy Selection.

# A CONCEPTUAL FRAMEWORK FOR THE DESCRIPTION AND ANALYSIS OF MAN-MACHINE SYSTEM INTERACTION

ERIK HOLLNAGEL *)

The paper presents an approach to the description and analysis of man-machine interaction, based on cognitive systems engineering, which focuses on the distinction between the operator's model of the machine, and the machine's image of the operator. The purpose of the cognitive task analysis is to achieve the model-image match that is the basis for efficient man-machine interaction. This is done through aligning control requirements with cognitive function characteristics. A particular example of a man-machine system model is introduced, and its use in the evaluation of system changes is shown.

## THE INADEQUACY OF ENGINEERING ANALYSES

System analysis is the necessary basis for Man-Machine System (MMS) design, which must consider not only the physical functions of the system, i.e. the mechanical or engineering aspects, but also the control functions and the interface between man and machine. It is possible to make an engineering analysis of the machine, and define the functions it must perform to fulfill its purpose. But for an engineering analysis to be sufficient for the whole system would require that the human was completely under the control of or subsumed by the machine. This is not possible with complex systems which generally involve process control rather than manipulation. Furthermore, the interface (which includes the physical interface as well as procedures for retrieving information and exercising control) cannot be designed from a purely engineering basis. To do so would imply that the operator could be described exhaustively and adequately in physicalistic terms, i.e. as a machine. Every experience so far shows that this assumption is fundamentally wrong, cf. Hollnagel (1).

The traditional approach has been to analyse the machine, determine where demands to its operation would exceed human capacity, and modify its design on those points. The operator's functions have thus only been considered as limiting conditions for the machine. Consequently, the maximal efforts of the operator have often been necessary to meet the minimal

*) Institute of Energy Technology, Halden, Norway.

requirements of the machine. Such a Procustean approach is fundamentally incorrect, Taylor and Garvey (2), since the proper functioning of such a system requires that the operator is maximally efficient. Rather than having the human adapt to the machine, one should adapt the machine to the human, so that human functioning has the best possibilities, i.e. so that the maximal requirements of the machine can be met with the minimal efforts of the operator. Accordingly the analysis of the operator's functions must address the performance characteristics and the cognitive bases for human performance. The latter cannot be derived from the engineering analysis of the functional requirements of the machine, i.e. the purely mechanical description, because psychological facts obviously cannot be deduced from engineering facts (1).

## THE SYSTEMS APPROACH

The necessity of integrating engineering analyses with ergonomic or human factor analysis in system design has been recognised for some time, at least by ergonomists and human factor specialists.

> Since the operator is an integral part of the MMS . . . the functions he performs must also be developed as a part of overall system design. Behavioural development may proceed in the absence of any specific ergonomic efforts, but it occurs even so, as a fallout from whatever engineering decisions are made. Hence, it is incumbent upon the systems ergonomist . . . to control the development of these behavioural functions with just as much conscious deliberation as hardware development is controlled. (Meister (3)).

This has, however, not involved a reconsideration of the basis for human factors engineering. The criteria for effectiveness are still dictated by the requirements of the machine. "(The operator's) limitations are taken into account to prevent the system from demanding more of him than he can provide; but in all other ways his functions are constrained to optimize system outputs" (3). In this view the operator is considered a part of the machine (and describable in mechanistic terms), rather than as one cognitive system interacting with another in the control of a process.

## MODELS AND IMAGES

Cognitive task analysis, Woods and Hollnagel (4), is concerned with analysing and describing the cognitive functions of a particular MMS. There are two parallel but interacting themes in this process. One is the operator's model of the machine, the other is the machine's image of the operator. Based on training, experience, instructions, and the nature of the interface, operators develop a set of internal models that describe the structure, function, relationships and operations of the machine.

These models are used to guide data collection, for planning and decision-making, and to evaluate the effects of actions. One requirement for effective human performance in MMSs is that the operator's models accurately capture the relevant properties of the system.

A logical extension of this idea is that artifacts also include a model or image of their environment and the user. The image of the user's physical characteristics is generally implied by the way in which the machine functions rather than explicitly stated. Practically every industrial product may be used as an example, and the inherent image becomes more crucial for the proper application, the more complex the function is. Similarly, designers build into their systems assumptions about the user's cognitive characteristics. A typical example is that interfaces, whether a washing machine, a telephone, or a computer terminal, require the operator to remember certain kinds of data and follow certain rules; in other words, there is an implicit assumption about how the user's memory functions and how performance is organised. Mismatches between the machine's image of the user and actual user cognitive characteristics (or performance characteristics in general), is a major source of system malfunctioning, generally attributed to 'human error.'

> Forcing people to interact on the machine's terms is not only inconvenient -- more importantly, because it is an unnatural mode of interaction, it is a primary cause of human error. (Norman (5)).

Therefore, a second requirement for successful human performance in MMSs is that the designer incorporates a realistic prototypical image of the operator's cognitive functions into the system design.

## The Model - Image Match

The relationship between the user's model of the machine and the machine's image of the user points out that overall system performance is not merely the sum of the performance of its human and machine components. Rather, there is a more complex interaction where design decisions within the machine domain affect the requirements to the cognitive functions of the human. This is recognised in cognitive systems engineering which changes the referent for the term "system" from the machine element alone to the machine / human elements together. Cognitive task analysis (CTA) is a way of analysing the functions of a man-machine system that explicitly acknowledges the importance of attaining a match between the operator's model and the machine's image. It provides a description of the tasks the system must perform to accomplish its function, as well as the principles that can be used for allocating subtasks among the parts of the system.

## CONTROL FUNCTIONS AND CONTROL REQUIREMENTS

One pole of the CTA process begins with the objectives of the machine. This refers to the designer's description of the machine's function. This description may be given in several modes, e.g. as a traditional engineering description or by an analytical principle such as multilevel flow modelling. The description usually takes the form of control functions, i.e. what the MMS must accomplish for the machine to meet its objectives. The control requirements of the machine may be specified from an analysis of the physical characteristics and purposes of the machine alone, disregarding knowledge of the operator. They must be fulfilled no matter how the control task is carried out and no matter how the control system is composed. Functional analysis is used to provide a top-down decomposition of the control function structure and relationships. The specification of the control function structure provides a way by which the activities required to accomplish each function can be identified. It thereby also defines the functional limits for the operator's model of the machine.

From a description of the control requirements, it is possible to derive a tentative prescription of the activities that are needed by the MMS. A separation into activities assigned to the computer and the operator respectively, can be achieved through task allocation principles. For the operator, it is further possible to describe not only the expected types of activity, but also whether an activity can be expected to be a skill, or something that requires attention and consideration. This must of course be done with the experienced operator in mind since the novice tends to rely less on skills than the expert. The actual design of the MMS, not least the operator support, must take into consideration what the "normal population" of operators will be like.

## COGNITIVE FUNCTION CHARACTERISTICS

The cognitive analysis is concerned with the capacities of the MMS to fulfil the control requirements. It is therefore concerned with the general functional characteristics of the MMS as a cognitive system, i.e. the domain specific cognitive functions. The activities of an MMS are ordered or organised, whether they are initiated by the operator or the computer, since they are in both cases planned by man. It is important to know how actions are organised and executed, because it sets the limits for human performance in the cognitive sense. A candidate description is provided by the skill-, rule-, knowledge framework, Rasmussen (6), which focuses on classes of decisions in process control and supervision. It must, however, be augmented to include the feedback from actions as well as a recursive decision structure in order to be generally applicable for a cognitive analysis.

Human activity can be characterised by means of response modes. Empirical data show that people tend to function on the level of automatic action, i.e. skills. The reason for that is in the nature of the human cognitive system. Attention is strictly limited, both in span and duration. If we only can do what we can attend to, we can do very little, e.g. Reason and Mycielska (7). This is clearly shown by the development of children, which is characterised by the gradual mastering of new skills. To begin with walking in itself requires full attention to the exclusion of everything else. Later on walking becomes a skill, hence something that can serve other purposes (fetching, bringing, moving). The cognitive function characteristics deal with the modes of functioning of man, in particular the relation between these modes and human performance variability or 'human error'. The description is on a far more detailed level than the general modes of data driven and concept driven performance, and concentrates on the specific rather than the overall organisation. This description sets parts of the functional limits for the machine's image of the operator.

## Cognitive Aspects of Man-Machine Allocation

The key concept in the CTA approach to man-machine allocation is that the tasks to be performed can change as a function of allocation choices, cf. Hollnagel and Woods (8) for a detailed discussion of this observation). This means that the allocation process is a part of the cognitive task analysis iterative spiral rather than a subsequent or isolated activity.

Given some allocation decision the resulting cognitive task structure must be analysed. If an undesirable consequence is the result, the initial decision is re-evaluated and alternatives to it are explored; otherwise, the initial decision is identified as a workable allocation. In other words, this approach expands the basis for allocation decisions to include the cognitive consequences of particular allocation decisions. There are a variety of factors that can influence a particular allocation decision or the choice of a general philosophy about man-machine roles in a system, including cost, available technology, and control philosophy among others. Cognitive task analysis does not invalidate these considerations; rather, it adds the point that there are varying consequences in the cognitive domain from allocation decisions and that these consequences affect user tasks, interface requirements, and ultimately the allocation decisions, thereby closing the loop. This mutual dependency contributes to attaining the model - image match.

## AN EXAMPLE: THE MAN-MACHINE SYSTEM IN A NUCLEAR POWER PLANT

The following sections will give an example of how the principles outlined above have been used to produce a model. We shall look at the domain of Nuclear Power Plants, and show how the Man-machine interaction can be analysed and described by means of a Man-Machine Process Control System (MMPCS) model.

## Types of Targets

First of all, we must specify the types of targets or goals that are relevant for assessing the function of a Nuclear Power Plant (NPP). This corresponds to the level of functional purposes in the abstraction hierarchy described by Rasmussen (9). Nuclear Power Plants generally have three distinct but interdependent targets which, in alphabetical order, are: (1) availability, (2) efficiency, and (3) safety. The targets are all high level abstractions, usually given in terms of the impact of the NPP state upon some external parameter.

AVAILABILITY means that subsystems and components are ready when needed. The target of availability implies maximising speed of response and minimising deviations of important variables. Availability is, of course, only important if the NPP production is needed. Otherwise it ceases being a target.

EFFICIENCY means that the process takes place as economically as possible, both in the short and the long run. Efficiency obviously depends on an external demand for the product. It is closely related to profitability and is in a sense the ultimate target, since NPPs presumably are built because they are believed to provide the most efficient way of producing electrical energy.

SAFETY means that the process does not inflict damage on people (the public or operators) or objects. For NPPs this refers particularly to uncontrolled release of radiation in any form. One must, however, not forget that other types of damage can result from the process if physical or chemical energy is released in an uncontrolled way.

The targets can be related in various ways, depending on the dimension of description. The relations will indicate how changes in one target lead to changes in the others, provided the environment is stable. One example of the relations on the dimension of Importance, is shown in Figure 1. A '+' means that changes occur in the same direction in two targets, a '-' that they occur in opposite directions.

## Types of Tasks

The targets specify the overall purposes of the functions of the NPP. Changes in the MMPCS that jeopardise the attainment of any of the targets, or which facilitate one target at the cost of another, are clearly unacceptable. In order to assess this a more detailed description of the basic tasks of the system is required. By a task is meant the sequence of activities that must be carried out by the MMPCS to achieve a target (or a subgoal on the way to the target). In the terms of the previously mentioned abstraction hierarchy, tasks are on the level of generalised functions.

The set of basic tasks for a NPP can be derived either analytically, for instance by use of flow models, Lind (10), or through empirical studies. An example of the latter is an outline of the tasks in an NPP as state space diagrams, which show the possible states of the process and the transitions required to move from one state to another, NKA/KRU (11). Although state space diagrams are graphic representations they may also be formulated as text strings that can be directly input to computers, Jacob (12).

A state space diagram may, of course, be made as detailed as desired. However, as the level of detail increases, so does the dependence on a particular process or plant. For the purpose of an MMPCS model it is desirable to have a common state space diagram for a general Light Water Reactor NPP, as shown in Figure 2.

The states in the state space diagram correspond to the goals that must be fulfilled to achieve the targets. While most transitions correspond to tasks, a few do not. Since tasks are the set of activities or functions that must be carried out to bring the system to a specific state, transitions that take the plant from a normal to a disturbed or emergency condition cannot be tasks, as it cannot be a goal to bring the plant into a disturbance.

Types of Functions

The next step in the development of the MMPCS model is to relate the tasks to the functions of the actual MMS. Each configuration of tasks and targets puts specific requirements to the MMPCS, generally in the form of control functions that must be fulfilled. The configuration thus defines a context for the MMPCS. On this level we can specify the functions of the MMPCS required to accomplish the target / task configuration, although we cannot yet specify the functions of agents, i.e. the operators and computers. The MMPCS is still considered as a whole. In terms of the abstraction hierarchy we remain at the level of generalised functions, although they refer to a different domain than in the case of the tasks.

General Process Control Functions. An example of a list of general process control functions is the following set of task categories given by Rouse (13).

| | |
|---|---|
| *COMMUNICATING* | *PROBLEM SOLVING* |
| *COORDINATING TASKS* | *RECOGNISING* |
| *EXECUTING PROCEDURES* | *RECORDING* |
| *MAINTAINING* | *REGULATING* |
| *PLANNING* | *SCANNING* |
| | *STEERING* |

Since this list only contains functions or task categories for which individual models are available in some form, it is necessarily incomplete. Furthermore, some of the items are of little interest for NPP process control, although they may be essential for other types of dynamic processes (e.g. steering).

The functions mentioned above are also very different in scope. Scanning, for instance, is a very specific activity for which there are a number of detailed models. Planning, on the other hand, is a much broader activity, which is difficult to model and which also requires a different type of models. Furthermore, there are some functions that are missing entirely, for instance Predicting and Identifying. A revision of the list of general process control functions is therefore necessary before it can be used for the MMPCS model.

Process Control as Information Processing. Another approach is to consider the functions in terms of the processing of information. There are several ways in which the functions can be described as information processing, again based on either analytical and empirical approaches. A generally accepted description, which resembles a decision making model (9), divides the functions of the MMPCS into the following steps:

INFORMATION COLLECTION. This is the continuous scanning and collection of data from the input channels (sensors) to the MMPCS, as well as the retrieval of more specific pieces of information.

INFORMATION ANALYSIS. The information analysis may be either part of the general monitoring of the system, or serve of one of the functions below. It is generally quite unsophisticated, in contrast to the functions of status identification and evaluation.

STATUS IDENTIFICATION. The analysed information is used to identify the status of the system. Normally the system is monitored to check whether or not it deviates from the target. If so, status identification becomes more specific in an attempt to characterise the off-normal condition.

STATUS EVALUATION. This relates mostly to off-normal conditions, where the MMPCS must evaluate the current system status to assess the expected consequences of action alternatives, in preparation of an actual action decision.

ACTION DECISION. Based on the status evaluation, the MMPCS must choose one of the appropriate action alternatives. Action decision may also involve the development of strategies for action, and prediction of the outcome of possible actions.

ACTION IMPLEMENTATION. Once the decision is made, the action alternative must be implemented. This may require further planning or breaking down of the action into its constituents.

ACTION MONITORING. It is necessary that the plant is monitored to see whether the intended conditions obtain (feedback). This may lead to the specification of further information collection, thereby closing the circle.

Compared with the list of general process control functions the information processing account is more homogenous in the sense that the functions are approximately of the same scope and on the same level of description.

The Process Control Functions. In order to reach a suitable list of functions it is necessary to extend the information processing list with a few more which are on the same level of description. What is missing is the capacity to retain the experiences from controlling the process (recording and learning), and to share information with other systems (communicating). When these are added we end with the following set of functions, ordered alphabetically.

| | |
|---|---|
| *ACTION DECISION* | *INFORMATION COLLECTION* |
| *ACTION IMPLEMENTATION* | *INFORMATION ANALYSIS* |
| *ACTION MONITORING* | *RECORDING* |
| *COMMUNICATING* | *STATUS EVALUATION* |
| | *STATUS IDENTIFICATION* |

These functions cover the essentials for an MMPCS. The logical next step is to describe the principles which specify the combination of functions for a task, and the distribution between agents.

## Allocation Strategies

The description of how the functions can be allocated between the agents of the MMPCS, i.e. the operators and computers, is independent of the task and function descriptions, and must provide the basic facts about capacities and limitations of the agents for the various process control functions. The success of a computer-based process control system depends strongly upon the allocation of the functions among the operator and the computer. It is particularly important to recognise what impact the use of computers for some functions has on those handled by the operator. An important consideration is that the operator at all times is kept well informed about the status of the process and the plant, including the effects of the tasks and functions handled by the computer. This implies that one should not aim for automation of as many functions as possible, and leave the rest to the operator. Instead one should distribute them according to a principle of functionality, and in compliance with the operator's inherent capacity to serve as an active agent in an MMPCS.

Until recently, the prevailing attitude was that function allocation started from a consideration of what could adequately be mechanised or automated, hence assigned to the computer. From an ergonomical point of view, the opposite approach would obviously be preferable. The best solution is, however, to consider the capacities and limitations of both agents, and allocate the functions from that basis. The assessment of agent capacities must logically precede the allocation of functions. This requires a thorough knowledge of human capacities, limitations and preferences, in addition to knowledge about machines and computers. Unfortunately, human capacity can vary considerably due to factors such as stress, anxiety, shifting working hours, in addition to acute demands for quick and rational action.

Even without knowing the capacities in detail, there is an obvious difference in how the strenghts and weaknesses of the agents are distributed. The important point here is to consider how a change introduced in a function or a task may propagate through the functional network. Since there is a limit to the capacities of the agents, both humans and computers, any change introduced in the system will disrupt the current balance and result in a changed distribution of agent capacities over functions. That may necessitate a change in function allocation, hence a change in how other tasks are carried out. This latter change may be latent, i.e. it may not show itself under the circumstances considered for the new design, but only appear under e.g. disturbed or accident conditions. A good MMPCS model should make it possible to predict (and prevent) that.

The principles for an allocation strategy completes the description of the main concepts of the MMPCS model. The four concept categories (targets, tasks, functions, and allocation strategies) correspond to four levels of description which are necessary to account for the dynamics of an MMPCS. Although each of them accounts for an important aspect of the MMPCS, none is more essential than the others, and none can be used alone.

An account of the ordering of the many factors that have an influence on the functioning of the MMPCS is given in Figure 3. They are grouped into three categories which we may call primary, secondary and tertiary. The MMPCS model must adequately account for the functional relationship among the concepts in the first category. It must also go into considerable detail about the influence of the factors in the second category. And it must recognise the potential influence of the tertiary factors, although it need not go into any detail regarding them. A different starting point would have resulted in a different structure. But basically the model should contain the same factors, although with emphasis on different aspects and relations.

## How to Use the Man-Machine Process Control System Model

There are several uses of the MMPCS model, both related to changes in man-machine systems. The primary is to assess the consequences of a specific change, particularly whether it achieves its stated purpose and whether it will have any unexpected side effects. A (suggested) change to an MMPCS is assumed to have a specific purpose, so that the expected effects on the system can be described in detail. These can relate to all of the main concepts in the model, i.e. a target, a task, a function, or an allocation strategy. The purpose of the change may, for instance, be the facilitation of recovering to normal after disturbed operation.

We can ascertain the effects of the change by relating it to the functional network of the MMPCS model. On the level of tasks it may, for instance, increase the likelihood that the NPP goes from a state of disturbed operation to a state of normal operation, rather than to a state of emergency operation. That will also have an effect on the targets. Similarly, the facilitation of a task implies a facilitation of the functions of the task. Since every function is part of other tasks as well, we can use the model to estimate what the consequences will be for other tasks, hence for the target / task structure as a whole. The evaluation of a given change, can be summarised by a series of questions that should always be asked.

The first question concerns WHERE a change is aimed, i.e. whether it is related to a target, a task, a function, or an allocation strategy. The answer must clearly identify which part of the MMPCS is involved, and how general / specific the change is supposed to be.

The second question concerns WHAT the purpose of the change is. The answer must specify the intended effect of the change on the functioning of the system. Depending on which level is involved, it could be a change in transition probability, a facilitation of a specific task or function, or a changed distribution between agents.

The third questions concerns HOW the change is intended to effect the function of the MMPCS on that particular level. The answer must identify the parts of the functional network that are involved and the way in which they are supposed to function differently after the change.

The fourth question concerns WHAT the possible side effects of the change are. These can either be on the same level as the main effects or on another level. A change in a task may, for instance, have consequences on the level of functions. It is a safe assumption that not all side effects have been anticipated. The usefulness of the MMPCS model should be obvious in answering this question, since the detached view of the model may compensate for optimistic design assumptions.

The fifth question concerns HOW the effect can best be measured. The answer must be a definition of the most sensitive measures for the change. This is crucial, since it is futile to make any attempt to assess the effect of a change unless proper measurements are defined. This question is closely related to the problem of validating the function of a system.

In addition to this, a series of more specific questions may be generated, which depend on the particulars of the change. The purpose of the MMPCS model is to guide the questions and organise the answers. In the present stage of development this is done manually, but it is conceivable that the process could be automated by implementing the MMPCS model as a computerised expert system. In any event, the use of models such as this will improve the consistency of descriptions and analyses of man-machine interaction.

## REFERENCES

1. Hollnagel, E., 1983, Int. J. Man-Machine Studies 8, 135-143.

2. Taylor, F. V., and Garvey, W. D., 1959, Ergonomics 2, 187-194.

3. Meister, D., 1973, Ergonomics 16(3), 267-280.

4. Woods, D. D., and Hollnagel, E., 1984 "Cognitive task analysis", in preparation.

5. Norman, D. A., 1980 "Errors in human performance", San Diego, California, Centre for Human Information Processing.

6. Rasmussen, J., 1979 "On the structure of knowledge - a morphology of mental models in a man-machine system

context", Risø-M-2192, Risø National Laboratory, Roskilde, Denmark.

7. Reason, J., and Mycielska, K., 1982 "Absent-minded? The psychology of mental lapses and everyday errors", Prentice-Hall, Englewood Cliffs, New Jersey.

8. Hollnagel, E., and Woods, D. D., 1983, Int. J. Man-Machine Studies 18, 583-600.

9. Rasmussen, J., 1984, "Strategies for state identification and diagnosis in supervisory control tasks, and design of computer based support systems", In W. B. Rouse (Ed.), Advances in man-machine systems research. Vol. I. J.A.I. Press, Inc.

10. Lind, M., 1982 "The use of flow models for design of plant operating procedures", Risø-M-2341, Risø National Laboratory, Roskilde, Denmark.

11. NKA/KRU-(81)12, 1981 "Enlarged Nordic cooperative program on nuclear safety. NKA/KRU project on operator training, control room design and human reliability: Technical report on operator training", Technical Research Centre of Finland, Helsinki, Finland.

12. Jacob, R. J. K., 1983, Communications of the ACM 26(4), 259-264.

13. Rouse, W. B., 1981, ACM Computing Survey 13(1), 71-99.

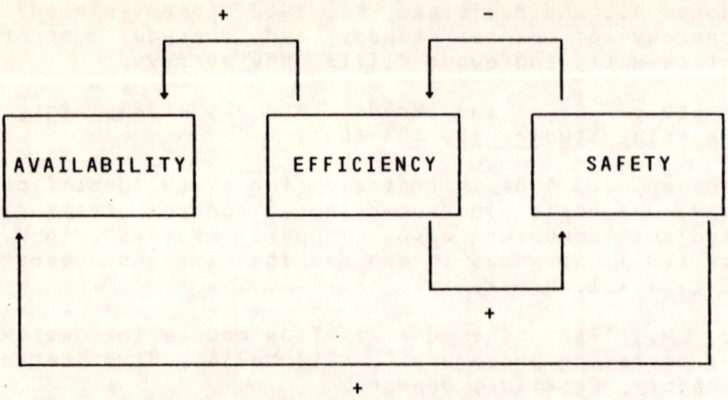


Figure 1: Relations Between NPP Targets on the Dimension of Importance.

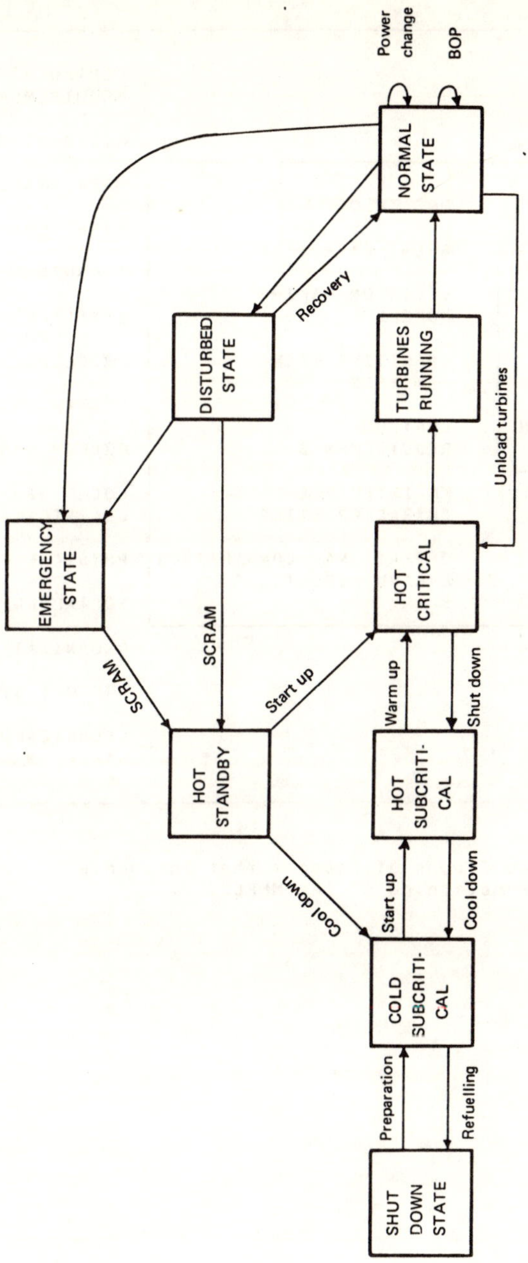
Power change
BOP
NORMAL STATE
Recovery
DISTURBED STATE
TURBINES RUNNING
Unload turbines
EMERGENCY STATE
HOT CRITICAL
SCRAM
SCRAM
Start up
Warm up
Shut down
HOT STANDBY
HOT SUBCRITI-CAL
Cool down
Start up
Cool down
COLD SUBCRITI-CAL
Preparation
Refuelling
SHUT DOWN STATE

FIGURE 2.

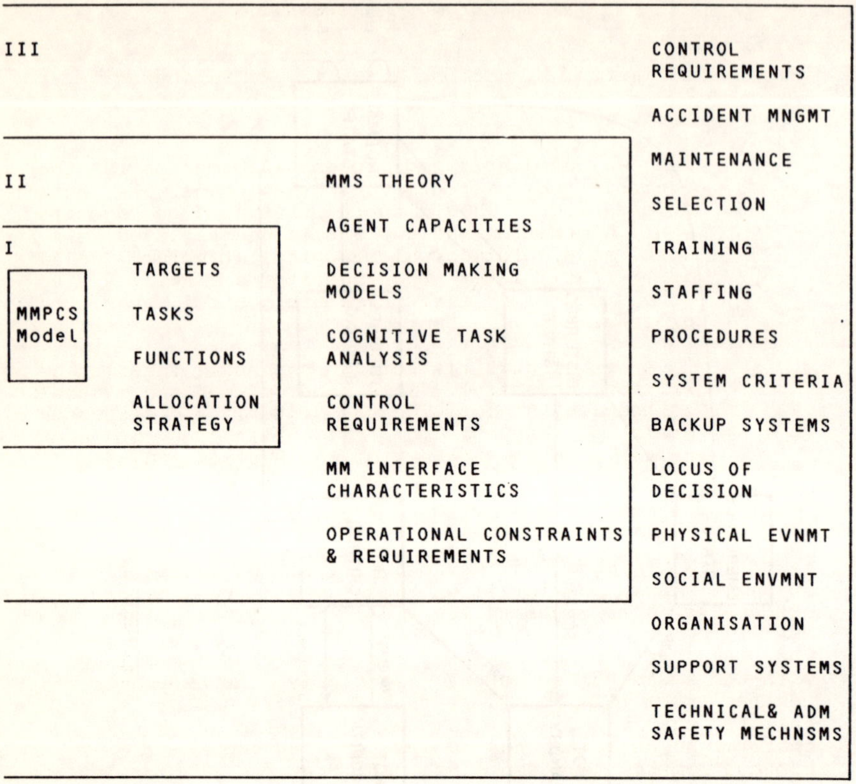


Figure 3: Three Groups of Factors that Influence the Functioning of the MMPCS.

# THE ROLE OF VISUAL INFORMATION IN IMPROVING THE ACQUISITION OF ADEQUATE WORK ORIENTATION

A. Kautto*, L. Norros*, J. Ranta*, H. Heimbürger*

The way of displaying process information in the control room effects the formation of the work orientation. Thus, there is a need to specify what information and how should be displayed, e.g. what mode, form or hierarchical structure would best facilitate the operator's ability to understand and exert flexible control over the process.

A set of criteria for use of visual information will be necessary. This paper presents a general methodology of deriving the criteria and discusses practical aspects for the realisation of such criteria in the context of nuclear power plant.

## INTRODUCTION

Along with the changing technology, i.e. the interactive display systems, new possibilities have been created for signal processing, data presentation and control functions.

The role of process operator and supervisor has changed along with the changing technology, especially with the video based control room (VCR). There is a clear trend that the operator has to supervise more complex processes than before. An operator's decision has a greater significance in terms of process safety, availability, efficiency, maintainability and product quality. Decision making requires problem solving activities at the abstract level of knowledge and a deep understanding of process operation principles. The increasing automation level also means that the operator might to a wider extent be faced with disturbances outside his experience. In diagnosing a disturbance the operator uses his knowledge of the process and also the immediate process information available to him. The need for preprocessing, structuring, ordering and presentation of the basic process information and dynamical data is of great importance.

The VCR is an important topic in the area of control room design. The VCR is nowadays based on cathode ray tubes (CRT). The VCR has been selected as a topic for this paper because there is

* Technical Research Centre of Finland, Electrical Engineering Laboratory, Otakaari 5 I, SF-02150 Espoo.

- an increasing interest in VCR design
- great flexibility for displaying and illustrating plant information for operators
- a need to solve problems using VCR, e.g.:
  - -- how to realize control of components
  - -- how could the limited displays (2...5 % of the total information) be used in an efficient way.

A pressurized water reactor (PWR) has been selected as an example of a complicated process.

This paper concentrates on disturbances, because the operators might have problems in handling the displayed information in complicated transients. In addition safety has been selected as a sub-goal for this study.

The following general questions can be raised:

a) What kind of a work orientation is needed in the new control activities and how could it be developed in an efficient way with the help of the interactive display system ?

b) How could the concepts and the complex interactions of the process be learned with the help of the interactive display system ?

c) How could new alarm functions be illustrated effectively by using computer graphics ?

d) How could the operator be supported in analysis of a disturbance by using the interactive display system and computer graphics ?

e) How to organize a display hierarchy according to the process hierarchy and how can the design information (sub-goals and design arguments) be included succesfully in the display system ?

## PROBLEMS IN CONTROLLING A COMPLEX TECHNOLOGICAL PROCESS AND SOME MEASURES FOR IMPROVING THE ACTIVITY

Working activities have traditionally been learnt on the job, whilst carrying out the required tasks. The knowledge of the process has thus been formed through empirical generalizations of the specific features of the process and includes associations between the features and certain operative measures. This is the nature of the individuals' process feeling, the quality of which differs from one worker to another. The knowledge of the process is typically of the senso-motoric or perceptual type and is integrated with operative skills.

Yet, the growing complexity of the technological process and the corresponding working process breaks the integration between knowledge and skills mainly because the knowledge needed in the control of the activity exceeds the limits of individual experiences, and because in the automated processes there is a reduction in the practical activity and thus also a reduction in the accumulation of experience that could serve as basis for empirical generalizations. There are many manifestations of the separation of these two facets of activity in the practical working situations. It is also formulated in the psychological theory and experiment. As an example we could mention Broadbent and Aston (1978) (7) who found that verbal knowledge of the process and the ability to control it are relatively isolated. We claim that the separation of knowledge and skill and also the variations in the quality of process feeling are a result of a spontaneous learning process but are by no means inevitable. Moreover, the problems in the acquisition of the process feeling are often said to be a result of the abstractness of the process information in the modern automated plants. Our interpretation would be that the more mediated (or abstract) information emphasizes the separation of knowledge and skill but is not the essence of the disintegration of the activity.

Knowledge and skill were said to be the two facets of activity referring to its cognitive and practical (causing change) functions. The crucial point in learning to control complex processes is to find the forms of activity that produce the knowledge and skill needed in the new conditions. In todays's technological environment stages the activity is basically internal , it is thinking. In the following, we make an attempt to explain our approach in developing the process operator's activities.

We use the term work orientation to represent the structure of knowledge and internal activities that form the basis for proper working activity in process control tasks. This concept includes both the cognitive and motivational aspects of activity. One could say that the so called process feeling is a spontaneous form of the work orientation that has to be consciously developed in todays working activities. From the work of some prominent theoreticians in the field of cognitive psychology (Galperin, Talyzina, Bruner) one could find some criteria for assessing the levels of work orientation a person can have. On this basis Engeström (1982) (1) defines three assessing criteria or dimensions: form of appearance, degree of organization and depth of explanation of the knowledge. The stages along these dimensions are seen in Figure 1.

Achievements along these dimensions mean steps away from every day spontaneous thinking and towards theoretical reflection of the objective world. In process control there are many concrete single indications of the objective pressure to refine the work orientation in the described way (for example development of functional oriented display systems and experienced need for design information in operating the system). It is the researcher's lot to make the correct generalizations of these indications and to try to direct and strengthen the tendency.

One cannot pour knowledge in anybodys head. Knowledge lives only in active thinking. Modelling the object helps in the transition from spontaneous to theoretical thinking. There are several advantages of visual models in developing the adequate thinking. Creating models allow the operator to make thought experiments on the object. This mental operation on the object is the only way of acquiring knowledge of it. This activity forms a reflective attitude toward the object and also toward ones' own thinking. It is an essential feature of theoretical thinking and it demands intrinsic work motivation of the operators.

The operators should be motivated and taught not only to solve the actual operating problems but also to strive at general understanding of the process. It is to note that both kinds of problem orientation appear spontaneously in different problem solving situtations, (Galperin) (8).

A further advantage of the visual modelling of the object is an easier comprehension of the interactions of the object and thus developing of a systemic approach is promoted. It is important to emphasize that we do not see the role of visual models as to illustrate the otherwise too "abstract" information but instead have them as a means of creating that activity in which the theoretical understanding of the object can be achieved. Theoretical understanding should not be confused with "something that is learned from books" but rather it is defined through the highest values on the three dimensions in figure 1. A further important aspect of theoretical thinking which has relevance in the process control context is that it allows generalizations on the basis of just one case.

In this paper we present tentative results of our conceptualization of the PWR process. With the aim of helping the operators both to enhance the comprehension of the process and to achieve efficient control skills, we have tried to explain the principles of deriving the displays from the essential physical nature of the process. Figures 2 and 3 are examples of the most fundamental models of the process that are refined further in the display design. Our conviction is that the learning of the operator activities involves much the same process as the design of the process and its information system. A systematic assimilation of the concepts in that way allows the development of a work orientation that is based on scientific and not everyday concepts. The assimilation demands active learning activities. The further development of didactic methods for that is the next step in our study on modelling and acquisition of work orientation.

## DECOMPOSITION OF INFORMATION

Information can be decomposed in many different ways. The goal is to find such a decomposition criterion, that will lead to a meaningful decomposition. This means that the operator is able to control the whole process with a few pieces of information at a time. Such a decomposition criterion is:

> Decompose the entity into the parts such that the common information of any pairs of the parts is at a minimum.

A process generally consists of refinement stages and transfers of intermediate products. A complicated refinement stage is a system, if its common information with a rest of the process is at a minimum. Transfer of the intermediate products forms a system only if it includes major auxiliary functions.

A lower level decomposition criterion of information is a classification according to the different technical means of maintaining the process functions. These task groups are auxiliary tasks supporting the main functions of the process, automatics, controllers etc., (the lowest level in Figure 6).

The information system must represent the process one to one. Then the information system will help the operator to get the right work orientation about the process. It follows that the operator will get the answer to the questions: Why the function or subsystem has been done (purpose), what can be done and how, Figure 4.

In addition to the preceding decompositions there exist the decomposition based on functional sub-goals. At first the necessary and sufficient conditions, subfunctions, must be studied in order to fulfill the sub-goals. The final step is to check when the subfunctions exist in the process, Figure 5. The figure 5 is on the more concrete level than the Figure 4.

One could imagine that the developing of the work orientation can be made faster, if the decomposition of the information can produce the result, that one (sub)function belongs to one (sub)system. In fact this could be consciously stated as a design criterion for a process.

As an example, the decomposition of the PWR process according to the sub-goals safety, availability and location will be presented. First the decomposition criteria must be formulated. This is presented in table 1.

Table 1. Criteria of the decomposition of the information concerning to the sub-goals safety, availability and the process location.

| Sub-goal | Criterion of decomposition |
|---|---|
| - safety | prevent a radioactive release fom core |
| - availability | a system's working implies its conditions to be fulfilled (to produce power) |
| - process | minimize mutual interactions (= common information) between any pair of subsystems |

In the following stage the meaning of the decomposition criteria is to be interpreted from the point of view of process state and control. In this example, the decomposition criterion of the safety sub-goal "prevent of radioactive release" can be interpreted such that mass and energy balances (m- E-balance) must be maintained within the desing conditions. E-balance is directly related to the efficiency of heat transfer. The decomposition of the information presented in table 2 represents the whole process from the point of view of sub-goals safety, availability and location. Also the use of the terminology has been defined in table 2.

Table 2. A general decomposition of sub-goals safety, availability and process (location), Kautto (3).

| | Critical Safety | Critical Availability | Process Location |
|---|---|---|---|
| Decom-position | - main functions (m- & E-balance heat transfer)<br>- functions<br>- subfunctions<br>- variables | - main components' function<br>- subsystems' function<br>- auxiliary systems'f.<br>- sub aux. systems'f.<br>- help level including automatics, controls etc. | - systems<br>- subsystems<br>- auxiliary syst.<br>- sub aux. syst. |

## PRESENTATION OF INFORMATION

### Design of displays

Design is a question of the composition of single displays in such a way that the operator can easily control the process. This aim can be achieved by dividing the process into the following dimensions and different abstraction levels, table 3.

The phrase "disturbance or normal event" in the lower right corner of the table 3 reflects the necessity of an efficient alarm function in the sense that the human should be able to handle the presented information. This implies that the number of simultaneous alarms demanding control actions should be minimized. This in turn implies the successful design of automation.

The leading principle in display composition should be, that "The entity should repeat itself in its parts", Lomov (4).

Table 3. The illustrate interpretations of the process dimensions.

| Dimensions of the process | Levels of abstraction | | |
|---|---|---|---|
| | General level | Intermediate level | Detailed level |
| - time | - dynamics | - gradient | - event |
| - location | - systems | - subsystems | - components |
| - functions | - m- & E balances and heat trasfer | - interactions between subsystems | - disturbance or normal event ? |

This can be interpreted as particular requirements for information presented on the displays. At least the following information should be available.

1. The state of functions (with the help of the alarm system).
2. The most important interactions between the (sub)systems.
3. The design information of the (sub)systems.
4. The location of the present display in the total display system.

It seems that the requirements 1, 2 and 3 will lead to a hierarchical display system. Also Figure 5 and table 2 anticipated this. Of course one can ask what other meaningful ways exist to present information.

A problem of the large display system as compared to the fixed information presentation in the conventional control room is that the operator faces an additional mental load in accessing individual displays, Mitchell (6). In order to avoid this additional load the following requirements can be stated.

A. The need to change displays should be minimized.
B. Accessing a display must be as easy as possible.
C. Individual control operations should be combined into actions of higher hierarchical level.

The requirement A is consistent with the preceding requirements 1, 2 and 3. Further the requirement A can be intrepreted as follows:

A1. Maximize the amount of information on the single display.
A2. Part of the information will be displayed only, when
   - something occurs
   - on operator's request.
A3. Same information must be partly included on the different displays, if needed.
A4. Dynamical information should exist on the displays in order to minimize the use of paper documents. Moreover the most

important design information should be included on the displays.

The interpretation A1 implies the need for on effective alarm function so that the operator would not have too much information (but enough) for decision making. There is a need to display both symbols and analog information because man easily understands diagrams in a "parallel mode". Alphanumeric data are handled in a "serial mode". Alphanumeric data are needed because of accuracy. A way of amending alphanumeric data into qualitative information is to use colourcoding e.g. normal, cautionary and alarm colour.

The requirement B can be interpreted as follows:

B1. The access of displays should also be bound to follow the logical, easily learnable identification system.
B2. For a hierarchical display system, access should be available:
- forwards or backwards
- up in hierarchy
- return to the top display of hierarchy
- to the previous display.
B3. The communication interface must be responsive to the operator's request.

The requirement C can be interpreted as follows:

C1. Serial control sequences should be amended to a single control in order to save time.

The preceding interpretations mean that both different usages of the display system and different ways of controlling it are feasible. However, the interpretations are simple and natural, and they guarantee a flexible use of the display system. It is feasible to define more complicated, chained accesses, but they would only be additional to the primary accesses stated in B2.

An example

System level displays. The decomposition of information and the coupling of single displays to each other in order to get the logical and natural access of displays leads to the structure of information presented in Figure 6.

The process and instrumentation (PI) display is the most important display for monitoring and control. Displays should logically reflect the process. In a power process energy is generated, refined and transferred in the process but water circulates in closed loops. The loops and energy transfer are usually presented as shown in Figure 7. This model can be the contextual basis of the display system. Energy is generated on the left side and it will be refined in moving to the right. These main stages, called systems, will be placed to the system level, table 2 and Figure 6.

Figure 8 presents the technical model of the process on the system level. This is derived from the principal model in Figure 7. The system level displays include only main components.

Subsystem level displays. By completing the technical flow diagram with subsystems and auxiliary systems a more complete presentation (Figure 9) will be achieved. Displays on the subsystem level are derived by cutting down the completed technical flow diagram of Figure 9 in certain sections. The subsystem level displays include components. The logical presentation of information implies also, that the direction of water flow on the subsystem level must be consistent with that of the system level displays; steam flow from the left to the right and condensate from the right to the left.

In order to fulfill the requirements listed in chapter 3 the single display area is divided into the subareas presented in Figure 10.

Individual control operations will be done on the subsystem level. The control of automatics could be realized on the system level because they represent a more general level of process management.

Selection and use of colours. Colours have to be used economically and logically. Pastel colours should be used in the normal state. The use of colours should be consistent with the population stereotype. The use of colours in our project has been presented in table 4.

Table 4. The use of colours on dark background.

| Colour | purpose |
|---|---|
| bright blue | steam phase and components |
| dark blue | water phase and components |
| violet | components in electrical system |
| bright green | normal dynamical in formation for steam |
| dark green | " " " " for water |
| blue green | " " " " for electricity |
| yellow | warning (availability) |
| red | alarm (safety) |
| half intensity yellow | support to return the failed availability function for use |
| pink | support to return the process to safe state |
| brown | component in maintenance |
| half intensity brown | fixed text |

A specific example

A PI display according to the requirements presented above is shown in Figure 11. It is called "turbine by-pass" and it belongs to the secondary circuit. Further it belongs to the safety sub-goal. Figure 11 presents a simulated and imagined disturbance. The trip of the seawater pumps (VC) has interlocked the turbine by-pass valves in the close state. A simultaneous loss of coolant accident has caused heat transfer problem in the primary side. Reactor trip has caused the trip of generators as designed. As a final result the heat transfer is cut between the steam generators and the sea. This in turn has decreased the heat transfer ability of the steam generators. This is presented by a red curved line. The display shows also the alternative control strategies, as design information, to improve the heat transfer ability of the steam generators. The alternative strategies are indicated by changing the colour of the valves (pink).

## DISCUSSION

In this paper arguments have been presented, why the hierarchical display system is well suited for the monitoring of the power process. The design of the display system includes two stages. In the first stage the information will be technically decomposed into small groups. The selection of the decomposition criteria and the method of decomposition has an effect on how consistent the result of the decomposition is with the human ability to handle information. In the second stage, in designing the display system and single displays, human abilities must be considered, to ensure that the control of the process would be possible. With the limited number of simultaneous displays in the VCR, a powerful alarm function will be needed to find the information needed in a disturbance so that the operator could avoid the additional mental work load in seeking the necessary information.

Adequate management of the process implies a deep and clear understanding of the process and its phenomena. Above all it implies adding design information into the display system. In some cases the design information can be presented visually. The described display system could be one possibility of presenting necessary information. It is based on the following requirements:

- The entity will repeat itself in its parts
  - a powerful alarm function
  - the most important interactions between (sub)systems should be presented
  - the design information should be presented
  - the location of the present display in the display system should be presented
- Minimizing the need to change displays.
- Dynamic information should be included in the displays to minimize the use of paper documents.
- The grouping of the auxiliary functions according to the control tasks.

- The display system and its access methods should be based on the logical identification coding system of the process.

References


1. Engeström Y,: Perustietoa opetuksesta (Fundametal information about teaching). Valtion painatuskeskus, Helsinki, 1982. 172 p.

2. Heinonen R., Ranta J., Wahlström B., Computer aided design of control and safety systems - project plans. NKA/LIT 3.1(81)201. 7 p. (in Swedish)

3. Kautto A.M.T., How to illustrate the information of a complex process via displays. NKA/LIT - workshop on artificial intelligence in control of complex systems, April 7-8, 1983, Riso, Denmark. 11 p.

4. Lomov B.F.- Teoria, koe ja käytäntö psykologiassa (Theory, experiment and praxis in psychology). Kansankulttuuri Oy, Helsinki 1982. 353 p. (Original in Russian 1979).

5. Loviisa nuclear power station, Brochure, 20 p. Imatran Voima Oy, Malminkatu 16, Pl 138, 00101 Helsinki 10, Finland.

6. Mitchell C.M., Design strategies for computer based information displays in real-time control systems. Human Factors 25 (1983)4, pp. 353-369.

7. Broadbent, D.E., Aston B., Human Control of Simulated Economic System. Ergonomics 21(1978)12, pp. 1035-1043.

8. Galperin P.J., Kotik N.P., Luovan ajattelun psykologia (Psychology of creative thinking), Psykologia 1983:4, pp. 243-246 (Originally published in Voprosii psihologii 1982:5).


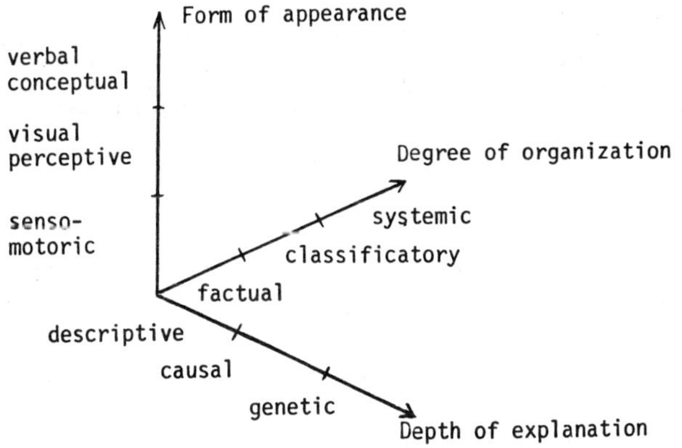


Figure 1 Dimensions of knowledge in the work orientation, (1, pp. 37, 39)

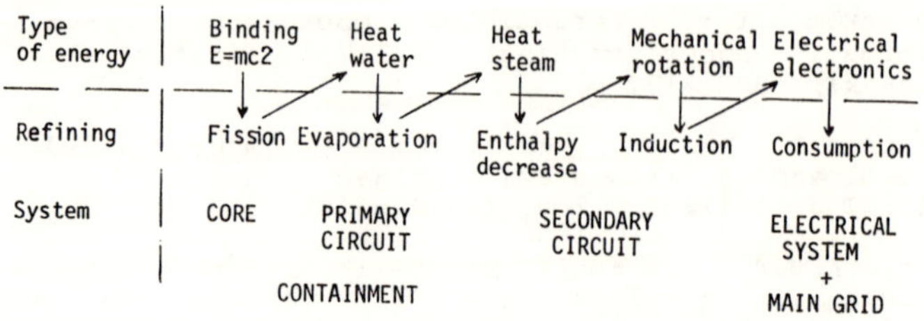


Figure 2 Fundamental physical phenomena in the power production and their classification according to the main systems

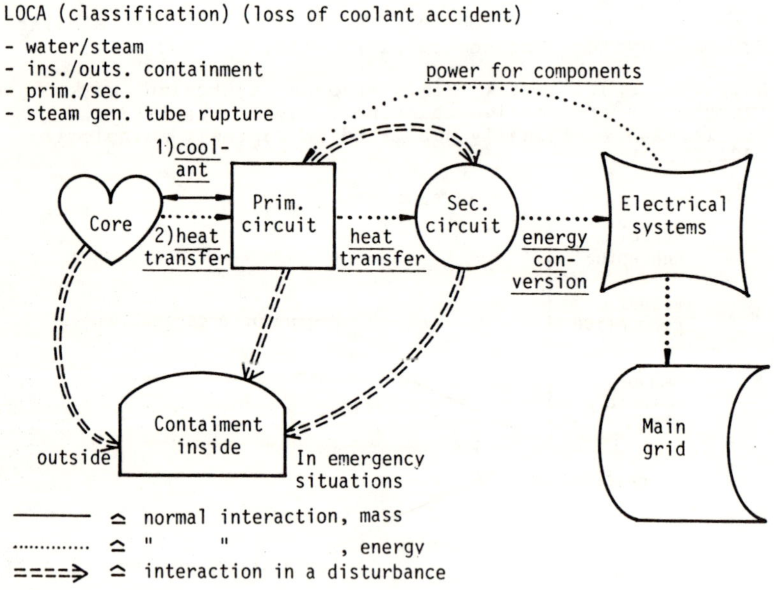


Figure 3 Interactions between systems

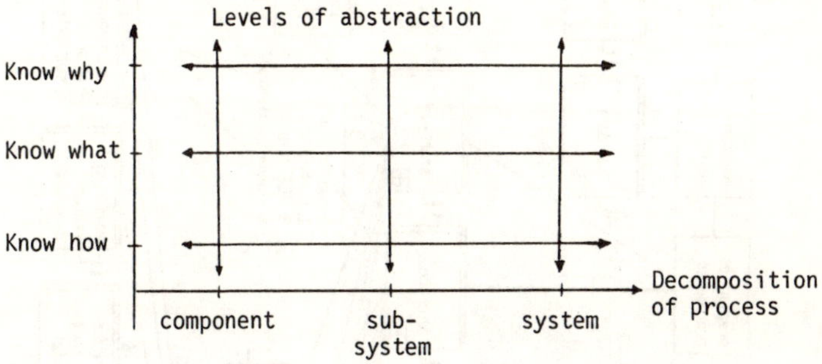


Figure 4 Levels of abstraction on the different process levels, Heinonen et al. (2)

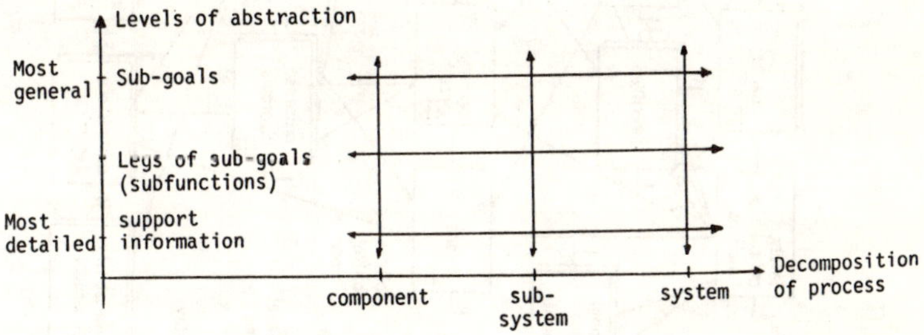


Figure 5 Location of the sub-goals on the different process levels

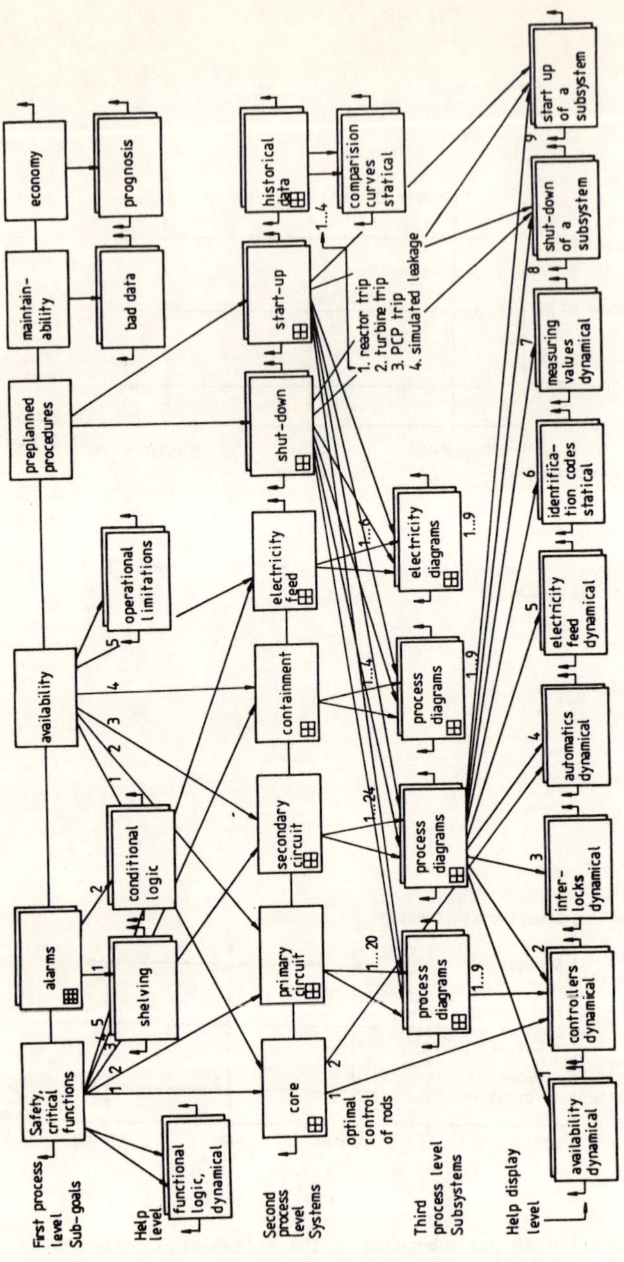


Figure 6 The structure of display system (3)

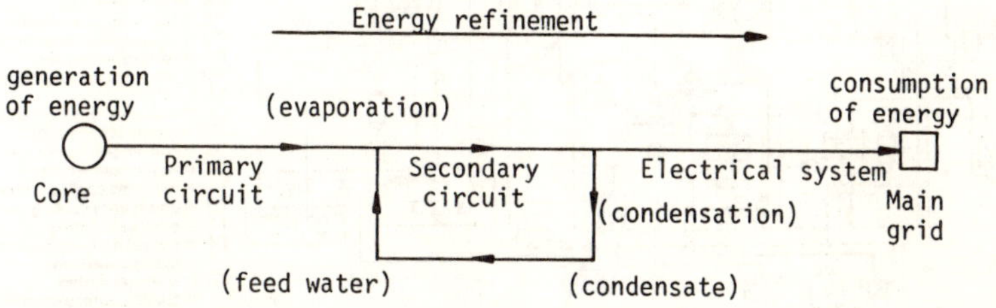


Figure 7 The principal flow diagram and energy refinement of the power process

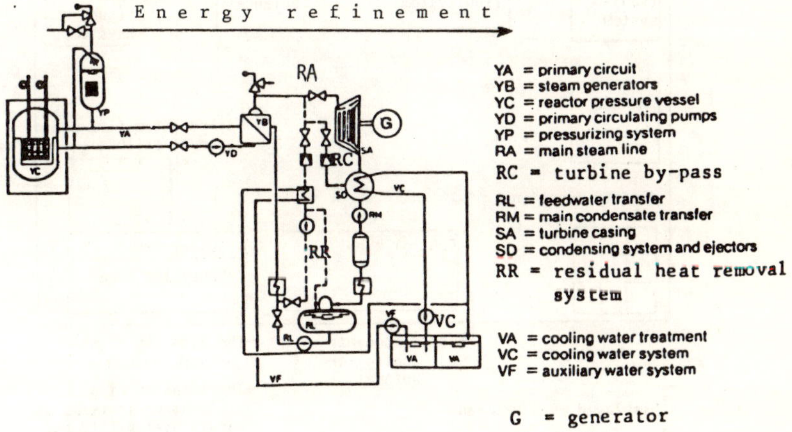


Figure 8 The main technical flow diagram of the power process (5)

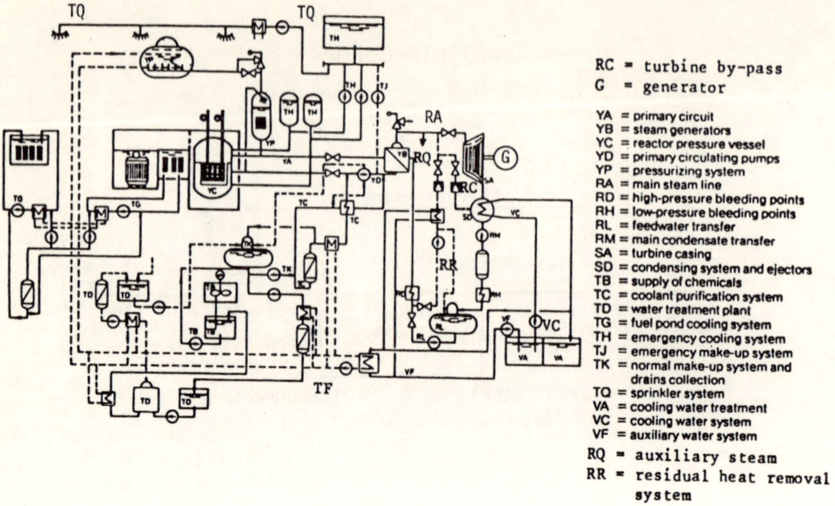


Figure 9 The completed technical flow diagram of the power process (5)

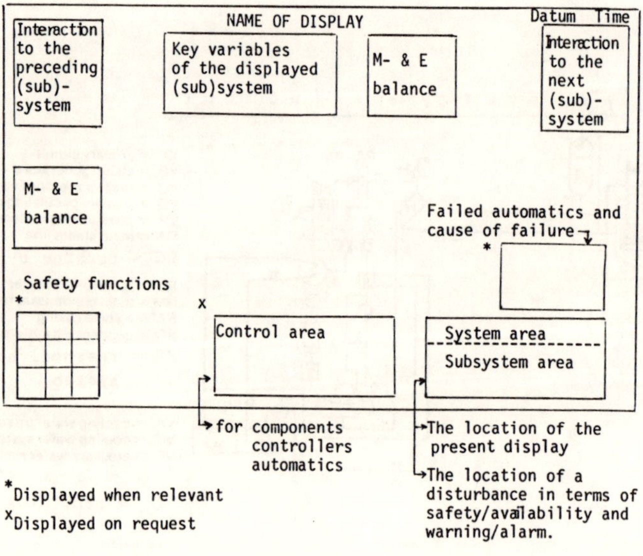


Figure 10 The structure of the single PI display

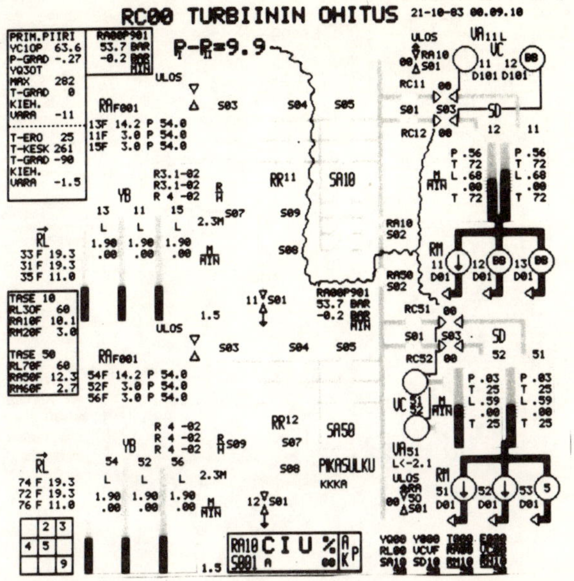


Figure 11 The turbine by-pass display of the secondary system, the third level of the display system (3)

# A COMPARATIVE SIMULATION STUDY OF ANNUNCIATOR SYSTEMS

H. Kragt*

In an experimental study, twenty-four trainee-operators from a chemical industry had to work with three different types of annunciator systems, based either on conventional or on modern instrumentation principles. Sequential information-presentation proved to be inferior to a system based on a simultaneous information-presentation both in performance and ratings.

## INTRODUCTION

The way in which information is presented to the human operator in control rooms of process industries is changing. Usually, the information was presented via control panels, but in the latest control rooms plant information has to be retrieved from Visual Display Units (VDUs). There are several reasons for this man-machine interface change. One is the wish to avoid the large size of conventional control-rooms. Another the need for flexibility which can be fulfilled by modern instrumentation. In conventional control-rooms the information is presented over whole wall surfaces (e.g. Seminara et al. (1)). Most of the time the operator gets the information just by walking along the panels. This way of information-presentation is called "parallel" or "simultaneous".

Dallimonti (2) suggests that the operator in those systems is not able to collect all the relevant information "at once", but instead has to "sample". To do so, it would be easier for him to sample from one position, via the VDU, the information he needs and to command the system which information he wants to be displayed. This way of information presentation is called "serial" or "sequential". Modern systems like TDC 2000, MICON, ICOSS, VIDEOSPEC, are all based on this principle.

Bergman (3) discusses this continuum of information accessibility and its related problems. Elsewhere, it is suggested that operators will experience difficulties in, for instance, fault finding, when plant information has to be retrieved from VDUs (Mazza (4); Marshall et al. (5)). Rijnsdorp et al. (6) mentioned the desirability of parallel presentation of information, instead of sequential presentation. In an earlier field study (Kragt et al. (7)), it was found that operators sometimes complain about the lack of a good overall view in the more advanced systems. For a better understanding of this problem and to trace systematically the pros and cons of both ways of information presen-

*Dept. of Indus. Engin., U. of Technology Eindhoven, The Netherlands

tation, a dynamic simulation experiment was conducted in which the effects were compared of simultaneous versus sequential information-presentation on performance.
From the results of the afore-mentioned field study, it was expected that information presented simultaneously would lead to a better performance than information presented sequentially.

## METHOD AND PROCEDURE

### Subjects

Twenty-four trainee-operators of a chemical industry in the Netherlands, between 19 and 27 years of age, participated. In three runs, each of which lasted 40 minutes, they had to work with three different types of annunciator systems. None of the subjects was familiar with any of these systems. They did not suffer from colour-blindness. Their training for the task lasted about 4 hours (pilot runs included).

### The simulated process

In the simulated process, a two component fluid (feed) is distilled into a "pure" top product (light component) and a bottom product (heavy component). The purity of both products was expressed in a quality measure ranging from 0 to 100. The subjects were instructed to hold the purity between 40 and 60 scale units ("alarm" limits) in order to make the products "saleable".
For a description of the equipment used, see Paternotte et al. (8) and Landeweerd et al. (9). For the present experiments the simulator was complemented by an annunciator system. The simulation model was adapted to the simulation of small disturbances in feed flow and feed concentration. Both disturbances influenced the afore-mentioned quality figures.

### Annunciator systems used

Conventionel Facia System (FS). This system presents the information simultaneously. The operator can see which signals are on and can relate them to each other. He can identify the light by its position. In the experiments, the operator had to acknowledge the signal by pressing the relevant button (individual acknowledgement). For a description of FS, see Kragt et al. (10).

Facia Picture (FP). As in the more advanced systems the information is mostly presented on a cathode ray tube (CRT), it was concluded necessary to exclude differences in visual aspects between lamps and CRT screens. Therefore a CRT picture was designed with the same structure as the facia system. One of the ways in which the facia picture system differed from the facia system was the way of acknowledgement. Typing a code was chosen because it would also be used in the advanced system and would exclude the differences in the way of acknowledgement (pressing a button versus typing a code).

Text system. The advanced system consisted of four "pictures". These pictures were presented one at a time on one CRT screen. The warning signals were time-order presented on the so-called "overview picture". The three other pictures were section pictures in which the warning signals had a fixed position. With the way of acknowledgement, "jumping" to a section and typing a code, sequential information-presentation was created.

### Task, objective and subjective measures

The following subtasks were selected: 1. Restarting pumps, after checking

conditions; 2. Controlling the process; 3. Making a written report; 4. Making an oral report.
For a detailed description of these subtasks (and the results), we refer to Kragt (11).

### Subjective measures

In addition to objective performance measures, interviews and ratings were used. Each subject had to carry out the task with all three systems. After each run the operator filled out a small questionnaire with regard to the system used, just to let him think about that particular system. After the three runs he was asked with an open-ended question "to compare the systems". Subsequently, the operator rated the three systems on:

- difficulty of restarting pumps;
- difficulty of controlling the process;
- difficulty of acknowledgement;
- overview of information presentation.

The scales involved seven equally spaced points ranging from 1 = very easy to 7 = very difficult. The operators were also asked to explain their ratings.

### Experimental design

It was expected that the differences between simultaneous (FS and FP) and sequential information-presentation (Text) would be greater if the signals followed each other more rapidly. In order to investigate a possible effect of signal density, two signal-density levels (CLUSTER and "EQUALLY-SPACED") were used.
The 3 annunciator systems (FS, FP and Text) which were investigated and the 2 signal density levels resulted in 6 conditions. A two-factor design involving 4 subjects in each of the 6 conditions was used. The 24 subjects were randomly assigned to the 6 conditions.
In order to compare the three systems, the same subject had also to work with the other two systems for the same density level. The sequence of annunciator systems was alternated.
To avoid effects of learning and fatigue, the data on performance of only the first runs were analyzed.

## RESULTS

### Restarting pumps

Mean pump starting-time. Analysis of variance (ANOVA) of the mean pump starting-times revealed a significant effect of the annunciator system ($F = 7.26$; $p<.05$) but no significant effect of signal density.
With the studentized range of Tukey (Miller (12)), we found that Text gave a significantly ($\alpha=.05$) higher mean pump starting-time than FS and FP. The latter two did not differ significantly from each other.

Pumps started too early. The number of pumps started too early was too small for further analysis. Remarkable, however, was the relatively great number of pumps (7 out of 24) started too early in the condition "Text CLUSTER" (all four operators being involved). Whereas the overall number of pumps started too early was 16 out of 144.

Difficulty of restarting pumps. The data were analysed with the Friedman Two-Way Layout Test on Rank Sums (Hollander et al. (13), p. 138). For the "equally-spaced" density, we found $S'=20.1$; for cluster: $S'=18.1$ (Critical value: $S'=6.167$; $\alpha=.05$). So, for both density levels there is a significant difference in ratings between the systems. Again a multiple comparison procedure was used to decide on the pairwise differences ((13), p. 151). For both density levels the operators rated Text as significantly ($\alpha$=.038) more difficult than FS and FP. No difference was found between FS and FP.

Overview of information presentation. For the "equally-spaced" density, we found S'=7.6; for cluster: S'=17.7 (Critical value: S'=6.167; $\alpha$=.05). So, there is a significant difference in ratings between the systems. The results of the pairwise comparisons showed that for both density levels Text was rated as more difficult to observe than FS and FP, the difference was statistically significant ($\alpha$=.038) for the cluster density only. No difference was found between FS and FP.

For both subjective measures the explanations given by the 24 operators were:

- "In Text you have to read the signals. Moreover the signals have no fixed position in the overview picture. On the contrary, on FS and FP you can see at a glance whether it is allowed to start the pump or not".
- "In Text it may be necessary to search for information. 'Jumping' to the sections and back to the overview picture is disadvantageous".

Conclusion. The sequential and alphanumerical way of information presentation and the lack of pattern recognition result in longer pump starting-times and unfavourable ratings for Text.

Controlling the process

Mean control-performance. ANOVA of the results revealed that the annunciator system had a significant effect (F = 9.27; $p<.05$) and that signal density again had no significant effect. With the studentized range of Tukey, we found that Text gave a significantly ($\alpha$=.05) higher score than FS and FP. The latter two did not differ significantly from each other.

Difficulty of controlling the process. For the "equally-spaced" density, we found S'=19.8; for cluster: S'=20.6 (Critical value: S'=6.167; $\alpha$=.05). This means a significant difference in ratings between the annunciator systems. Pairwise comparisons revealed that for both density levels the operators rate Text as significantly ($\alpha$=.038) more difficult that FS and FP. No difference was found between FS and FP.
The operators gave the following explanations for rating Text as the least favourable:

- "Too much time elapses before the control action can be carried out".
- "You have to read the signal before you know which one it is (no fixed position in the overview picture)".
- "Other signals are much more disturbing during controlling the process".

Conclusion. The control performance in FS and FP is better, because the subject knows immediately what he has to do. Processing a signal via FS or FP takes less time than via Text.

Written reports

From the analysis we learned that there were significant differences in number of errors between the annunciator systems. Less errors were made with FS than with FP and Text. The latter two systems did not differ in this respect. We assume, that the number of errors made is related to the way of acknowledgement. Text and FP take more time than FS. As a result, for the former two systems the present task had to be carried out faster.

Detailed analysis showed that more incomplete reported signals occurred with FP, whereas with Text more wrongly/unreported signals were found. Considering this, the question arises how the subjects rated the ways of acknowledgement.

Difficulty of acknowledgement. For the "equally-spaced" density, we found $S'=22.1$; for cluster: $S'=22.8$ (Critical value: $S'=6.167$; $\alpha=.05$). For both density levels the operators rated Text significantly more difficult than FS and FP (pairwise comparisons; $\alpha=.038$). No significant difference was found between FS and FP, but the values indicate that FS was rated as the easiest system in this respect. The explanations for the ratings were:

- "Dealing with FS, all you have to do is press the relevant button".
  Note: some operators did not prefer this system, because of becoming careless ("merely pressing a button").
- "Acknowledging a signal in the FP system actually means reading and typing a code, but with Text you also have to determine the relevant section and jump to it, and then you loose the overall view (much more difficult)".

Conclusion. It appears, that it is not the way of information presentation but the way of acknowledgement (typing a code) which affects the total number of errors made in the written reports. However, from the explanations of the ratings it was found that typing a code as such was not perceived as the main problem. "Jumping" to the relevant section in the Text system was found inconvenient.

## General conclusions

The results very consistently show that Text is inferior to both FS and FP. This result is due to the way the information was presented (sequential versus simultaneous) and not to the medium (hardware facia versus CRT). This conclusion is emphasized by the answers of the operators when asked which system they would advise for the present task; 13 operators preferred FS and 11 FP (simultaneous presentation). None of them preferred the sequential presentation.

In the present study, an annunciator system based on sequential information-presentation proved to be inferior to a system based on simultaneous information-presentation both in performance and ratings. In our opinion a simultaneous presentation is necessary for a task, like the one used in this experiment, where decisions have to be taken quickly.

The study is convincing in that new systems should be comparatively investigated using both performance measures and operator ratings. The latter in particular, because the operators are the ones who ultimately have to work with the system. In doing so, close co-operation between manufacturers, process industry and researchers, is necessary.

## REFERENCES

1. Seminara, J.L., Conzales, W.R. and Parsons, S.D., 1976, Human Factors Review of Nuclear Power Plant Control Room Design. Report EPRI-NP-309, Electric Power Research Institute, Palo Alto, California, USA.

2. Dallimonti, R., 1973, New designs for process control consoles. Instr. Technol., 20, 48-53, November.

3. Bergman, H., 1982, Design and evaluation of complex process-control systems. In European Research in Process Control: Human Interactions with Computer-based Systems (Edited by T.R. Edman and H. Bergman) (Proceedings of a Workshop presented at the 1982 Annual Meeting of the Human Factors Society, Seattle USA; Honewell-Technology Strategy Center,

Roseville, Minnesota, USA, October), section 2, pp. 1-19.

4. Mazza, J.D., 1977, A comparison of integrated and conventional cockpit warning systems. Master Thesis, Naval Postgraduate School, Monterey, California, USA.
5. Marshall, E.C., Duncan, K.D. and Baker, S.M., 1981, The role of withheld information in the training of a process plant fault diagnosis. Ergonomics, 24, 711-724.
6. Rijnsdorp, J.E. and Rouse, W.B., 1978, Design of man-machine interfaces in process control. In Proc. 5th IFAC/IFIP Internat. Conference on Digital Computer Application to Process Control (Edited by H.R. van Nauta Lemke and H.B. Verbruggen) (Amsterdam: North Holland), pp. 705-720.
7. Kragt, H., Timmer, C.M.J. and Delnoy, H.J.J., 1981, Private Communication.
8. Paternotte, P.H. and Verhagen, L.H.J.M., 1979, Human-operator research with a simulated distillation process, Ergonomics, 22, 19-28.
9. Landeweerd, J.A., Seegers, H.J.J.L. and Praagman, J., 1981, Effects of instruction, visual imagery and educational background on process control performance. Ergonomics, 24, 133-141.
10. Kragt, H. and Bonten, J., 1982, Evaluation of a conventional "process-alarm system" in a fertilizer plant. IEEE Transactions on Systems, Man, and Cybernetics, vol. SMC-13, no. 4, July/August, 1983, pp. 586-600.
11. Kragt, H., 1984, A comparative simulation study of annunciator systems. Submitted and accepted for publication in Ergonomics.
12. Miller, R.G., 1981, Simultaneous Statistical Inference (New York: Springer-Verlag).
13. Hollander, M. and Wolfe, D.A., 1973, Nonparametric Statistical Methods (New York: John Wiley & Sons).

# CONTROL CENTRES FOR NUCLEAR RESEARCH

B. Stumpe*

The European Laboratory for Particle Physics (CERN) in Geneva has a number of control centres from which a vast complex of accelerators are computer controlled 24 hours a day. Ten years' of experience with Touch Interactive Central Control has led to a new generation of modular compact multiprocessor assemblies for man-to-machine interfacing. The role of the keyboard is "secondary" in the sense that one tries to limit its role to the preparation of the interactive process (menu) where the touch screen is the primary tool.

## INTRODUCTION

The European Laboratory for Particle Physics in Geneva exploits a number of accelerators distributed over very large areas on the border of Switzerland and France.

Each accelerator is computer controlled from its own central control room. Although each of these control centres had been designed and upgraded by completely different teams, not necessarily having the same design objectives and constraints, a certain standardization of the physical interface between the operator and the "machine" has spontaneously developed with time.

This evolution started in 1971 when the central control system for the new Super Proton Synchrotron (SPS) was defined.

The basic design criteria for the central control was to multiplex via the computer network all information to one operator. The system should have a built-in feature of being able to convert process control which in the beginning could be manual (guided by the operator) into a fully automated process. As the main problem was the physical space needed for all the conventional buttons, knobs, and control panels a more flexible and less space demanding solution was required.

Inspired by work done by Fryberger and Johnson, experimenting with the idea of using touch interaction, replacing "hard" buttons with "soft" buttons, a touch panel was developed for use in the SPS. This panel allows up to 16 labels or buttons to be written under the touch screen.

* SPS Divison, C.E.R.N., Geneva, Switzerland.

The common denominator in all accelerator control rooms at CERN today is a man-machine interface using one or more touch panels as the primary tool for operator interaction. Associated with the touch panel is a number of input devices such as the tracker ball, knobs, shaft encoders, mechanical buttons and keyboards. The dominating output devices are TV based raster scan systems using black and white for touch screen interaction as well as for the keyboard echo and colour TV for the full graphics main screen.

## THE CONTROL PHILOSOPHY

The basic design criteria on which a man-to-machine interface should be built can be defined as follows :

a) "Operator friendly" which means :

- easy to use
- safe to use (the risk of doing the wrong thing must be minimal)
- fast in use.

Achieving all these things diminishes :

- the need for specially trained keyboard operators
- operator fatigue
- the need for additional training each time the system and its control possibilities are extended or reconfigurated

b) "Implementation friendly" which means :

- real time updating of all changes or additions to the control to be performed must be possible without any need to change the operator interface,

- there should be no limits to the total amount of operations possible (achieved by "soft buttons"),

- the human interface should be a universal controls tool being able to do any kind of control for any application and in any environment.

## THE CONTROL CONSOLES USED FOR THE CERN SUPER PROTON SYNCHROTRON (SPS)

The operator console is arranged such that one person can control all parameters of the machine from one seat without moving around. The space available for installation of displays and control devices is therefore extremely limited. It was therefore decided to reduce the number of input and output devices to a minimum but on the other hand to do the implementation in such a way that the various devices could be freely chosen by the computer for control of any parameters where an interactive process was requested.

For visualisation of the information the principle is to give to the operator only that part of information which is relevant and optimal in any particular point of operation.

Figure 2 shows the basic console components as they are used to control the SPS machine today. The following input devices can be seen :

a) The touch panel.
b) The computer controlled knob.
c) The tracker ball.
d) A few mechanical buttons.
e) The identification card reader.
f) The keyboard.
g) The card reader.

The output devices are the following :

h) Black and white TV monitor for the touch panel.
i) Black and white TV monitor for the keyboard.
j) Colour TV monitor for standard resolution (288 vertical x 384 horizontal).
Colour display for full graphics.
k) Monochrome TV monitor for double resolution (576 vert. x 768 horiz.). Full graphics display.
l) Alphanumeric display (16 characters for knob mode).
m) Loudspeaker for audio "click" or voice.
n) TV hard copy unit.

## TOUCH INTERACTION

As the main dialogue between the man and the machine is achieved using touch interaction something general should be said about this "ergonomic" method for operator communication with a given system.

One of the first things a small child tries to do is to point directly to (and if possible touch) the item it desires. Modern control systems which need a human operator in the loop take advantage of this primitive reflex when the "man-to-machine interface" is designed. The practical tool for its implementation is called the "touch terminal". Basically, therefore, a touch terminal gives full control to an operator who points his finger at one out of a menu of possibilities presented to him on a display. The device allowing a terminal to become touch sensitive is a transparent overlay to the display and is called a "touch panel".

### Current Techniques for Construction of Touch Panels

Today basically five different techniques are used for the construction of touch panels. These are :

- touch wires
- infrared light

- acoustic waves
- resistive
- capacitive

Which Technique Should One Prefer ?

Some systems have a relatively high touch resolution allowing an important number of "soft buttons" to be presented to the operator. Here one has to be extremely careful. A system presenting a large number of buttons might be, but is not necessarily better than a system working with fewer buttons.

One advantage of a high resolution system is that one has more freedom to place a small number of buttons merely where one wants them to be than where they are defined to be by the lower touch resolution.

What is most important is that the operator is presented with a limited amount of clearly perceptible choices on which he can take fast consecutive actions. A large number of buttons may require careful search (reading) of the screen causing unnecessary concentration and strain when trying to touch the correct button.

To give advice on which system to choose is more difficult. Basically all principles work, and work well. However, the best thing to do is to contact the manufacturer and get a list of implemented systems. The real users should then be contacted and answers to the following questions found :

- Does the touch panel transparency affect the readability of the display ?
- Do parallax errors affect the operation ?
- Is the touch properly debounced ?
- Are there any moving parts ?
- Does ambient light affect the system ?
- Can an insect start a process ?
- What effect does finger grease or scratches on the screen have ?
- What chemicals can be used if the screen has to be cleaned ?
- Is there an audio "click" as feedback to the operator when the button has been successfully pressed ?
- Is there any drift in the touch detection system ?

The CERN experience using touch screens is very positive. The touch panel developed for the SPS is now used in many places both inside and outside CERN.

The device which has been constructed uses a set of capacitors, etched into a film of copper on a sheet of glass and made of line elements sufficiently thin and well separated to be effectively invisible. Each capacitor is so constructed that the proximity of a flat conductor, such as the surface of a finger, increases its capacity by a fairly large factor. Figure 3 shows the transparency of the panel. Figure 4 shows a touch button in use.

Other devices using the optical method where a light beam is interrupted by a finger have also been built and tested at CERN. Some of the problems encountered were related to drift in sensitivity and ambient light problems. Other people outside CERN have had problems with insects flying into the light beam causing an unwanted action to take place. None of these optical CERN built systems has been put into use at CERN.

We believe that the main reason for the success of the capacitive system is the advantages of being purely electronic, having no moving parts or surfaces to corrode or wear out. It does not need a bulky frame around the screen. It leaves the display uncluttered and undedicated and the reliability is as good as that of conventional electronics. The screen construction is simple and therefore not expensive. A very much appreciated feature of this touch screen is the fact that a slight but recognizable finger pressure must be applied before any action takes place. The risk of accidental action when the finger is used for just pointing at the information on the screen is therefore strongly reduced.

## Future Applications

Touch panels are now being used in a large variety of applications. They are especially popular where very complicated systems have to be controlled rapidly and safely by a single person (accelerator and fusion control) but also where large data bases have to be consulted (health management systems, library access, etc.). One of the later applications for which systems are being developed, is traffic control. Systems for air traffic control already exist and systems for railway control have been experimented with.

Plans are now being made to introduce touch panels into the cockpit of tomorrow's military and passenger aeroplanes.

Touch panels are playing a more and more important role as an essential element of a modern man-to-machine interface. It allows unspecialized people to operate complex systems with a minimum of training.

However, the touch panel future (by number) is probably not in systems with substantial computer power behind them but in small systems which everyone uses every day. These can be washing machines, ticket automats, control stations in our own home, etc. where the microprocessor built into such system in the future needs a cheap and reliable standard interface for human interacton which even a child can use.

## OTHER INPUT DEVICES USED

The input devices on nearly all CERN consoles is limited to the types mentioned previously. The SPS machine uses a single multi-function computer controlled knob which together with the tracker ball has satisfied the need for control of variables.

This knob is very expensive and will therefore be replaced by the much simpler shaft encoder type of knobs which for many years has been used with success in the PS (Proton Synchrotron) and the LINAC control system. A new module called MMC (Man-Machine Communication) has been developed. The module which is controlled by an MC68000 processor looks at all the input devices previously named.

## The Role of The Keyboard

The primary role of the keyboard is in connection with the writing of the programs which are used to control the accelerator. The program language NODAL is an interpretive language, specially developed for the SPS with multi-computer systems in view, which makes it possible for technicians, physicists or others to write programs in a relatively clear language. The programs can if desired be executed step by step, and the computer system can also report back what has been executed in a clear language. Almost all programs can be executed by all the computers in the system.

Programs written in NODAL make extensive use of calls to data modules (or subroutines), each with their own function. These data modules are written in assembler language, and are stored in a memory local to each computer in the system. All the thousands of NODAL programs which are written for the running of the accelerator are centrally stored in a very big memory (disc) attached to the library computer alone.

The data modules are the only programs which are written by programmers, whereas the many NODAL programs are mostly written by "non-specialists", in other words the technicians and engineers who were involved in the machine's construction. About 20 of these are at present engaged full time for the operation of the machine, which runs 24 hours a day. The operators are working in three-man shifts of eight hours each.

The role of the keyboard is secondary during the actual control of the machine as normally all programs are called from the touch screen by which the operation is carried out. If wanted control actions can also be performed from the keyboard.

## THE CONSOLES FOR THE FUTURE

In 1981 the CERN Member States decided to add another accelerator to the existing complex of machines. This accelerator is called Large Electron-Positron (LEP) storage ring, which will allow intense beams of energies as high as 55 GeV. This machine is larger than the SPS, being installed in a 27 km long tunnel compared with the 7 kms of the SPS. The tunnel varies in depth between 40 and 70 m, and has eight access points.

As the two machines have to work in perfect synchronism (the SPS will be used part-time as injector for the LEP) it was natural to suggest that the central control for the two machines be done from only one main control room. Supported by arguments of economy concerning the needed equipment as well as the required total number of operational staff, it was decided to have one operations group only and to do the double control from the existing and now expanded SPS main control room.

## Experience With The Present Consoles

In a previous section describing the basic design criteria for a man-to-machine interface it was stated that:

> the human interface should be a universal controls tool allowing the operator to do any kind of control for any application and in any environment.

This is of course a very ambitious statement and such a perfectionism is difficult to achieve if not impossible. However, before going on with the design of new consoles for the future let us go back and have a look at the SPS consoles as they were designed ten years ago and how they look today.

It is interesting to observe that in spite of a continuous expansion of the number of machine elements in the system and a considerable increase in the variety and complexity of the equipment to be controlled, the basic components and to a large extent the physical appearance of the operator interface is still the same as no demands for changing its basic elements have been made. This does not mean that the consoles has not undergone any improvements, on the contrary ! Although the functionality of each element is still the same the computer power supporting these elements has been substantially increased. We can therefore conclude that the original design has proven itself and could be taken as a basis for the new generation of consoles.

## The New Requirements

The construction of LEP and the growing complexity of the SPS machine has triggered a number of requirements for new consoles. These can be listed as follows :

a) Console as a direct process controller (stand alone mode). The modules for control and test of the equipment is integrated into (part of) the console.

b) Console as a local area (central) consoles. The console is connected to the local area network and controls the local equipment through the network.

c) Consoles as central consoles for the entire machine The consoles are linked to the main communications network which in turn is linked to all the local area networks.

d) Console as a remote console but allowing access (and control?) like a central console. The console is linked to the local area network or main communications network going to a local area.

The last requirement is due to the large physical distances between the various buildings in the entire complex. The existence of such consoles at various geographical sites would allow a "specialist" working in a remote area or being at home to move to the nearest console when called through the telephone or the paging system. The permission to do this interaction should of course be given at central controls level only. An extension to this facility could be a portable console or a terminal allowing a person to make observations and maybe even controls from his own home.

## ELEMENTS OF THE FUTURE CONSOLE

Figure 6 shows the crate layout for a console which is composed of a number of modules where different processes for man-machine interaction can be controlled by a varying number of powerful microcomputer based (MC 68000) controllers. One module is normally a master, co-ordinating the bus traffic inside the crate. Each of the logical units has sufficient firmware to allow it to run a high level language in stand-alone mode and can be tested by connecting a terminal to the terminal port on the front panel. The global test of the assembly is done from the master in the same way using the same language.

This system offers the advantages of having each process separate which allows the future upgrading of one process only with a limited impact on the rest of the system. Sufficient slot space is available if new controllers are required later.

It should be noted that subsets are easy to produce if "maxi" consoles are not appropriate for a given task. A micro console could consist of only one colour monitor for full graphics overlayed with a touch panel.

Consoles of the type described (Fig. 6) are now fully operational and all the modules, apart from the communication interface to the various communication links, are commercially available. This includes the special software required for this universal type of man-to-machine interface. The first of these consoles are CAMAC based. Work is now going on to provide a functional equivalent in VME which has become the recommended standard for new CERN developments.

## CONCLUSION

Many years' experience has led to the construction of modular conceived multiprocessor consoles. The touch screen has proven to be a primary controls tool for any kind of equipment or industrial process. Man-machine interfaces which are really "ergonomic" can therefore be built as they are easy to use, application independent and therefore ready for new applications without expanding the physical space required for doing the operation from one seat only.

## REFERENCES


1. Fryberger, D., and Johnson, R., 1971, IEEE Trans. Nucl. Sci., NS-18, p. 414.

2. Suret, P., 1971, "Simplifying the man machine interface with the touch terminal", Software Division, Plessey Co.

3. Butler, H.L. et al, 1971, IEEE Trans. Nucl. Sci. NS-18, p. 419.

4. Beck, F., and Stumpe, B., 1973, "Two devices for the operator interface of the new CERN accelerator", CERN Yellow Report 73-6.

5. Crowley-Milling, M.C., 1975, "The design of the control system for the SPS", CERN Yellow Report 75-20.

6. Beck, F., 1976, "The design and construction of a control centre for the CERN SPS accelerator", CERN Internal Report, Ref. CERN/SPS/CO/76-1.

7. Crowley-Milling, M.C., 1977, New Scientists, 29.12, "How CERN broke the software barrier".

8. Crowley-Milling, M.C., and Shering, G., 1978, "The NODAL System for the SPS", CERN Yellow Report 78-07.

9. Anderssen, P.S. et al, 1981, CERN Internal Report, Ref. LEP Controls Note No. 11.

10. Hatton, V., 1983, "Man-machine interface versus full automation", Proceedings of Europhysics Conference on Computing in Accelerator Design and Operation, Berlin, Germany.

11. Shering, G., 1983, "Consoles and displays for accelerator operation", Proceedings of Europhysics Conference on Computing in Accelerator Design and Operation, Berlin, Germany.

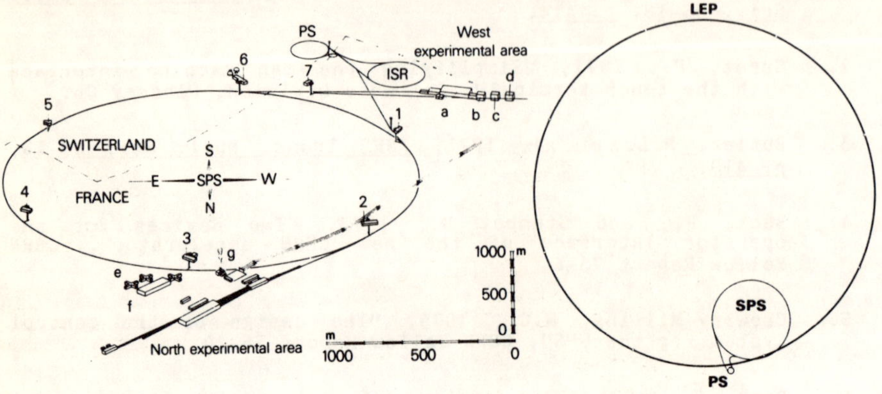


Figure 1 The CERN complex of present and future accelerators

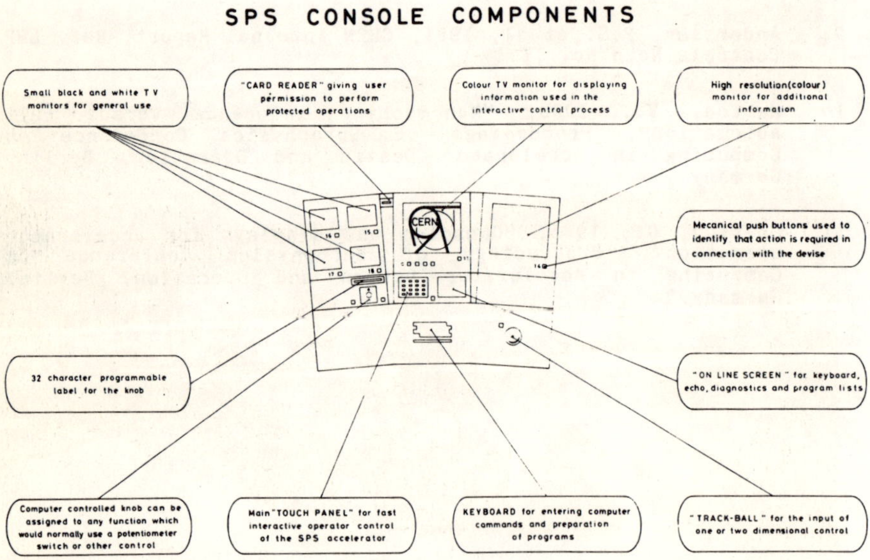


Figure 2 Basic SPS consoles

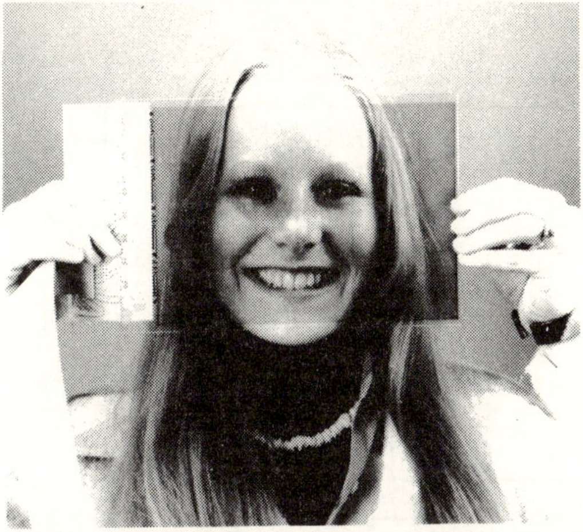

Figure 3 The SPS touch panel

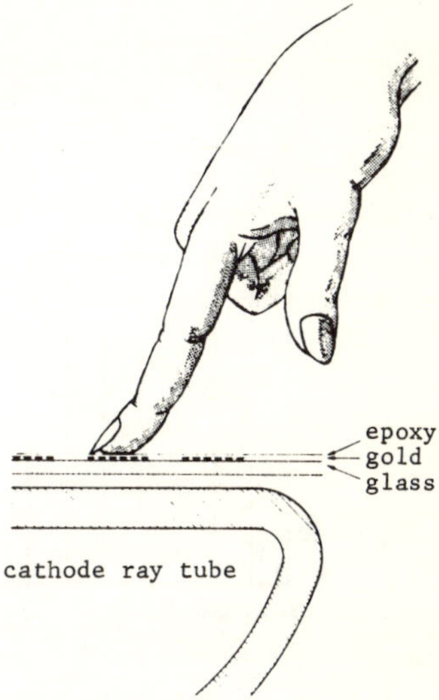


Figure 4 Touch button in use

Figure 5 Recent picture of the SPS control room (June 1983)

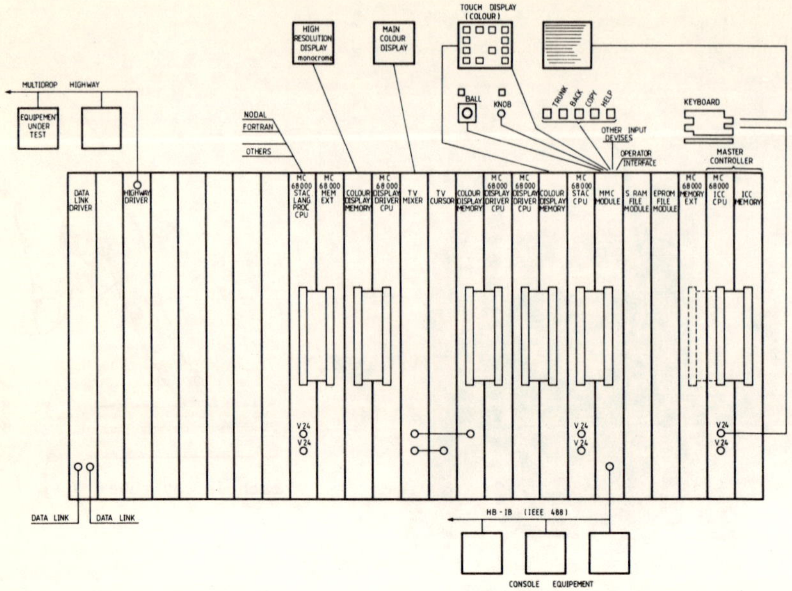


Fig. 6 Crate layout for a multiprocessor console

# EVALUATION OF INTEGRATED CONTROL AND SUPERVISION SYSTEMS

H.J.G. Zwaga* and M. Veldkamp**

A procedure for the evaluation of the human operator aspects of computerized process information and control systems is described. It consists of: (1) event recording, (2) observation of operator actions, and (3) interviews with operators. The procedure has been applied to two systems, with different levels of VDU implementation. The results have been set against the design principles nowadays adhered to by instrument manufacturers. The actual systems were found to present a number of considerable operating difficulties.

## INTRODUCTION

In the sixties digital computers were introduced in the process industries for control and data acquisition. Initially this resulted in hybrid control rooms: computer peripherals were installed next to the conventional panel instrumentation of indicators, recorders, controllers and switches.

Over the last decade more advanced computerized systems have been designed and installed, providing, apart from the initial data acquisition and data processing, also automatic control. In some chemical plants the availability of computer control is already of prime importance.

The justification of the use of these computers is usually threefold:

a. the system can provide better process and management information than conventional instrumentation,
b. the system provides more flexible control and optimisation of process conditions, and
c. the system can support the activities of process operators by providing data of better quality (higher accuracy, facilities for trend information on selected groups of variables, long-term averages) by generating warnings of operational limits being exceeded and by relieving the operator from relatively less rewarding, time consuming and sometimes less interesting tasks (control of, in general, slow parts of the process, load reduction of furnaces, etc.).

* University of Utrecht, The Netherlands.
** Koninklijke/Shell-Laboratorium, Amsterdam (Shell Research B.V.), Amsterdam, The Netherlands.

In view of the increasing pressure on the efficiency of operations (energy conservation, reduction of off-spec production and quality give-away) and quality requirements on one hand and the decreasing costs of computer hardware on the other, there is a general trend towards further computerization in the chemical process industry as a whole.

With respect to the functional specifications of these systems: those from the process technology and management side are mostly well understood and relatively easy to formulate, but the functional specifications of the interrelations between the computerized system and the human operator are far less obvious.

To gain some insight into this interrelation and to provide some guidelines for future system design, we have conducted a number of ergonomic evaluations of petrochemical plants with computerized instrumentation. A prerequisite for such an evaluation of a system is a clear insight into the design cirteria used in deciding on the operator's task and into the facilities provided by the man-machine interface. A wide variety of discussions with management and designers reveal that it is difficult for them to formulate these functional specifications.

The allocation of tasks between computer and operator, the work expected from the operator and the facilities provided in the operating console were studied by major instrument manufacturers and the resulting model of the operator's behaviour was used during the design of their new systems. A typical description of such a model is given by Fraser (1). When discussing the information display facilities of visual display unit (VDU)-based systems, he states: "The displays within such systems are generally arranged in an operational hierarchy. This is based on the way in which an operator works using a traditional control panel. The complete panel represents the entire plant, a section of the panel represents a process section and so on down to individual loop instruments. Under normal conditions the operator will simply want an overview of the plant, with no detailed information. If an alarm occurs or, indeed, if the operator requires more specific data, he will physically approach the appropriate section of the plant. He will then select a group of instruments and finally concentrate on one in particular. In a centralized display system a hierarchy of displays is provided to correspond to this procedure. The number of levels in the hierarchy may differ from system to system, but the principle is the same".

Reynolds (2) describes a study of operator behaviour in conventional control rooms and in control rooms with some form of computer interface. He also uses this model of operator behaviour when he speaks of the 'management by exception' rule. According to this rule the operator almost exclusively acts in reaction to external stimuli, mainly alarms. These design criteria, as formulated by two leading process instrumentation manufacturers, are also used by managements and designers of petrochemical production industries (Van der Meer (3)).

An important object of the evaluations presented in this paper is to validate the model used for the operator's behaviour, since this model has a direct impact on interface facilities for future systems.

## EVALUATION PROCEDURE

Other investigations conducted by us (unpublished) have demonstrated the usefulness of the procedure described on the next page. The information on the operator's behaviour collected for the evaluation is of three types.

(1) Event recording:

A standard facility in most computerized systems allows logging, in an event recording file, the occurrence and sequence of actions of the operator on the console. It has to be accepted that these data are in some way inaccurate, because other persons in the control room, such as technologists and maintenance engineers, also use the console facilities. We have assumed these non-operator events to be a constant error factor. As the event recording can be continued for a long period (weeks), we consider it useful for generating objective data on the relative frequencies of use of console facilities.

(2) Observation:

Three teams of two observers worked shifts of six hours on a 24-hour basis for five consecutive days. They coded activities of the panel operator in one of a number of predefined action categories (administration, looking at a VDU, control action, etc.). Definition of the action categories was such that all operator activities could be coded as one of a great number of types of actions distinguished for the investigation and stored on magnetic tape or floppy disk. The observers also recorded the onset of alarms generated by a hard-wired alarm system.

The observation data contain two types of errors. (a) A subjective element can seldom be avoided in the judgement of the observers. (b) The observation interferes with the operator's normal behaviour. The subjective element in the judgements was minimized by carefully describing the action categories and by organizing a number of training periods thought to be sufficient for the observers to learn the observation categories thoroughly. All training was done in the actual control room. In this way the training also diminished the effects of the presence of the observer on the behaviour of the operator. Observers and operators knew each other well before the start of the actual observation period. Furthermore, it was stressed at the announcement of the study that it had nothing to do with an evaluation of their personal performance, but was only aimed at an appraisal of the way the computerized system was used.

In both cases to be described later, a comparison was made between the frequencies of the events recording a week before, during and after the observation period. No significant differences were found.

(3) Interview:

The interview is an effective method of obtaining information about the subjective aspects of the task performance of the operator, such as his knowledge and experience of the process.

The aim of the personal interviews with a number of operators was to collect information about their working conditions, motivation and personal approach of the system. To steer the interviews, we made a list of topics to be discussed with the operators. The topics were selected in such a way that inferences could be made about the applicability of the design criteria used in the development of the system. Interviews were conducted with panel operators of all shifts and with some of the head operators. Most interviews lasted between 60 and 90 minutes.

The combination of event recording, observation and interviews enabled us to obtain an acceptably balanced picture of the operator's behaviour. It ranged from an objective registration of console actions (event recording) to a subjective appraisal of what the operator tells he is doing (interviews). The observation technique stands somewhere in the middle between the two other techniques: it provides an estimation of frequency and order of the actions of the operator not covered by the event recording and it allows a check on what the operator tells about the way he uses the system.

## CASE STUDIES

Two evaluation studies will be described. The first one deals with the evaluation of a conventionally instrumented plant extended with a computerized data acquisition and processing system. The man-machine interface of the system is part of the conventional control panel. The evaluated system, called PRIMA, provides int. al. target supervision and deviaton alarms and can be extended with computer set point control. The panel presents information both on parallel conventional instruments and on VDUs and can therefore be considered to offer a hybrid interface.

The second study related to the evaluation of an integrated control and supervision system, called ICOSS, in which the information is presented mainly on VDUs. ICOSS has no conventional instrumentation panel.

### The PRIMA Evaluation

System Description. The evaluation study was carried out at the propylene oxide section of the propylene oxide/styrene plant of Shell Chemie, Moerdijk. In this plant propylene and ethylbenzene are converted into propylene oxide (PO) and styrene monomer (SM). The plant consists of two sections: the SM-section and the PO-section, with the instrumentation of the PO-section housed in two consoles, each 13 metres long and facing each other. The system comprises 110 loops, has about 1000 non-controlled variables, including 300 derived variables calculated by the computer. At the time of the evaluation (1980) the system had been in operation for two years. Fig. 1 shows one of the conventional panels and a detail of the panel with PRIMA interface. It should be noted that the PRIMA interface is fully integrated in the conventional instrumentation panel. The PRIMA interface consists of: a matrix keyboard, an alphanumeric keyboard, a VDU, a numeric display and a printer (not shown in the figure). Three different kinds of information can be presented on the VDU:

(1) 'pictorials' : simplified flow diagrams of parts of the process with updated numeric information,

(2) 'all value report': provided are a numeric one-hour trend over the last 24 hours, a numeric 3-minute trend over the last hour and daily averages over the last two days of a single variable.

(3) 'specific reports': all other VDU reports, some of which are mainly intended for maintenance staff and technologists.

The additional information available from the numeric display is restricted to single measurements and is in engineering units. All information available on the VDU can also be requested as printed output.

The PRIMA interface does not enable the operator to enter any control action; all the operator's control actions are entered on the conventional panel. Only one operator is responsible for supervision and control on the panel.

Objectives of the Evaluation.

(1) Given the fact that the operator has a conventional panel, what kind of PRIMA information does he in fact use and what is the relative difference in use between conventional and PRIMA information?
(2) What kind of information does the operator use during normal and abnormal process conditions?
(3) What kind of information does the operator use after an equipment-in-danger alarm.

Results.

(1) The use of the conventional and the PRIMA information.
Events were recorded before, during and after the observation period of 120 hours. Total event recording time was 363 hours. Immediately after the observation period eight operators were interviewed. Owing to unavailability of the recording equipment, seven hours of the observation period could not be analysed.

Considering only the use of the conventional and PRIMA information sources, the observations showed that the operator predominantly used conventional information (about 75 % of his activities and 85 % of his time). Details of the frequency and the time distribution are given in Table 1.

The event recording indicated that request of PRIMA information was mainly restricted to 'numeric display' (64 %), 'all value report' (18 %) and 'pictorial display' (13 %).

It appeared from the interviews that the operators indeed consider the PRIMA information an addition to the conventional information. Advantages mentioned were: greater accuracy, availability of derived variables and the immediate availability of numerical data. The 'all value report' was said to provide a better overall picture of a section. It was mentioned that the 'pictorial displays' can be used for fault diagnosis.

In general, it can be concluded that the operator relies mostly on the conventionally presented information. PRIMA provides extra information and has some advantages over the conventionally presented information.

(2) Information used during normal and abnormal process conditions.
In consultation with plant management, the observation period was divided into periods with normal process conditions and periods with process upsets: the abnormal periods. Of the 113 hours covered, 74 are in normal periods and 39 in abnormal periods. During abnormal periods there is a relative decrease in the use of PRIMA information. Of the amount of time the operator collects information, he spends 92 % in front of the conventional panel. In the normal periods this is 80 %.

TABLE 1 - Absolute and relative Use of conventional Activities (1-4) and PRIMA Information Sources (Activities 5-9). The Total of Activities 1-9 (Consultation of Information Sources) is 100 %.

| Operator activities | Frequency | % | Time (s) | % |
|---|---|---|---|---|
| 1. Panel observation | 3150 | 37.9 | 44466 | 50.3 |
| 2. Flow scheme observation | 421 | 5.1 | 5133 | 5.8 |
| 3. Walking along console | 1228 | 14.8 | 15746 | 17.8 |
| 4. Control action | 1526 | 18.3 | 10444 | 11.8 |
| 5. Numeric display | 1284 | 15.4 | 4127 | 4.6 |
| 6. Specific reports | 51 | .6 | 538 | .6 |
| 7. All value report | 366 | 4.4 | 4485 | 5.1 |
| 8. Pictorial displays | 262 | 3.2 | 3109 | 3.5 |
| 9. Printer observation | 33 | .3 | 403 | .5 |

TABLE 2 - Relative Use of PRIMA Information Sources during normal and abnormal Periods.

| PRIMA-activities | Normal | | Abnormal | |
|---|---|---|---|---|
| | Frequency % | Time % | Frequency % | Time % |
| 5. Numeric display | 63.8 | 30.4 | 65.5 | 38.5 |
| 6. Specific reports | 3.5 | 5.7 | 0.3 | 0.2 |
| 7. All value report | 20.4 | 38.2 | 13.7 | 27.8 |
| 8. Pictorial display | 10.5 | 22.1 | 19.1 | 31.1 |
| 9. Printer | 1.8 | 3.5 | 1.3 | 2.4 |
| Total | 100.0 | 100.0 | 100.0 | 100.0 |

Table 2 presents a separate analysis of the PRIMA information actions during normal and abnormal periods. The results indicate that during abnormal periods the operator changes his use of the different information sources: there is an increased use of 'pictorial' and a decrease in the use of 'all value report'. The 'all value report', with its more general overall picture of a variable, is of less use than the 'pictorial' when the operator is diagnosing a disturbance and bringing the process back to normal, because the latter allows him to keep track of and compare the changes in a related group of measurements. This particular way of using the 'pictorial' was also suggested in the interviews. The 'all value report' is of more use when long-term changes have to be detected (process supervision).

(3) Information used after an equipment alarm.
During the 113 hours of the observation period 565 equipment alarms were registered, averaging 5 alarms per hour. In almost 40 observation hours no alarms occurred at all. There were 3 to 4 hours with oscillating alarms (29 to 62 alarms per hour), accounting for 30 % of the total number of alarms. As could be expected, most alarms were registered during abnormal periods (427). The averages for the normal and abnormal periods are respectively 1.9 and 10.9 per hour.

Analysis of the three-minute interval immediately after an alarm in abnormal and normal periods, shows typical differences between the amount of time an operator allocates to the different ways of collecting information. During normal periods he relies less on the consultation of others and on the use of conventional information and makes more use of PRIMA information than during abnormal periods (Table 3).

During normal periods an alarm is an incidental and isolated event. Many control actions with the accompanying panel observations and consultations will seldom be necessary. The operator will, however, want to verify that the alarm is not the start of a major disturbance. In this verification procedure the PRIMA system is used more during normal than during abnormal periods.

TABLE 3 - Amount of Time taken by three Groups of Activities during the three-minute Interval after an Equipment Danger Alarm has occurred during normal and abnormal Process Conditions.

| Activities | Process Condition | |
|---|---|---|
| | Normal (%) | Abnormal (%) |
| Conventional information | 30.0 | 37.5 |
| PRIMA information | 4.5 | 1.7 |
| Consulting | 22.5 | 30.6 |

Conclusions relating to PRIMA. Considering the amount of time, the regularity and the frequency of panel activities, it is clear that operators prefer conventional to VDU-presented information, especially during abnormal plant conditions.

The operators use the system in addition to their conventional information facilities. They appreciate the extra features of the information it can provide, such as: accuracy, and immediate availability of numerical data in engineering units. They have learned to use the specific PRIMA information 'all value report' and pictorial'. Overall, however, they rely mostly on the information provided by the conventional panel.

## The ICOSS evaluation

ICOSS (integrated control and supervision system) at Pernis refinery became operational in 1977. The system provides the integration of the information from, and control of, three high-vacuum units (HV) and one crude distiller (CD).

Two types of alarm can be generated by the system: (1) Hard-wired equipment danger alarms, generated independently of the computer by conventional alarm units. An alarm triggers an audio signal and an indication lamp starts flashing on an alarm supervision panel. (2) Performance alarms or deviation messages, generated by the supervision part of ICOSS when the values of one or more variables in a section of the plant fall outside predetermined ranges. These ranges are calculated from the target values for important process variables and depend on the mode of operation of the plant. When a deviation message becomes active, there is a bell signal and a group indication lamp in one of the alarm supervision panels starts flashing.

Fig. 2 presents a view of the ICOSS control room. There are two operating consoles: one for the HV-units and one for the CD-unit. On these consoles the operator can request the display of console-specific information and can communicate with the control system.

Fig. 3 shows the lay-out of the consoles. Access to the system for information display or control is via a graphic selection panel containing dedicated pushbuttons for control loop selection, or via a matrix keyboard for the selection of all variables and control loops by keying-in a tag number.

Above each of the operating consoles is a graphic alarm supervision panel. These panels display the process in a schematic form and contain the different warning lights mentioned above.

In front of the operating consoles is the supervision console. Supervisory information can be requested at this console; deviation alarm information, mode specification and special reports. There is, however, no control command function available in the console.

Adjacent to the supervision console is the message typewriter. It prints automatically all the deviation messages and can be used as a hard-copy unit for target specifications and special reports. Behind the supervision console is a bank of eight logging typewriters. They produce an hourly log of average process conditions.

The system has about 1200 selectable measurements, including about 100 derived measurements and 60 pictorial displays. The system comprises 220 loops (HV:112, CD:108). An extensive description of ICOSS is given by Bos et al.(4).

There are two different ways of selecting information on the consoles. Via the graphic selection panel there is direct access to all control loops. The information presented consists of three types of information. (1) Measured value, set value and valve output. All measured values are updated every 10 seconds. (2) Analogue indications of the above-mentioned values as a percentage of the scale 0 - 100 %. (3) A four-hour graphic trend of the variable selected, together with not more than three related trends (variables) on a VDU. On the matrix keyboard all variables and controllers are selectable via tag numbers. Information is either presented in the same way as for graphic selection or presented on the message VDU.

The pictorial displays present a special type of information. They consist of simplified flow diagrams or sections of a plant with a number of measured values superimposed. The display can also show alarm information. Updating takes place every 30 seconds. Pictorials can be requested via both the graphic and the matrix selection option.

Provided in the supervision console as well as in the operating consoles are 'group supervison buttons'. These buttons give direct access to deviation messages existing in a plant section. The messages are presented on the message VDU of either the supervision or the operating console, depending upon from where the request was made.

Objectives of the evaluation. When deciding on the control room lay-out, the information display and control facilities, the system designers assumed the following working procedure of the operator. In normal circumstances the operator sits behind the supervision console. When there is a deviation message he can determine the general location of the deviation on the alarm supervision panel above the operating console. The message typewriter provides him with point specific information. Via the group supervision buttons on the console he can easily request additional information, which will be presented on the message VDU in front of him. Once he has made his diagnosis, he can go to the operating console and make the changes needed.

When there is an equipment alarm he can determine its location from behind the supervision console on the alarm supervision panels, go to the operating console with the alarm and take remedial action.
The evaluation questions were:
(1) What is the working procedure of the operator and what kinds of information does he use.
(2) How does the operator react to equipment alarms and deviation messages?

Results.

(1) The working procedure used by the operator.

Events were registered during 341 hours, before, during and after the observation period of 120 hours. Immediately after the observation period eight panel operators were interviewed. Because of unavailability of the registration equipment, only 87 hours of the observation period could be used for analysis. Process conditions were normal during the observation period, no major disturbance occurred.

The most striking point in the event recording is the large number of 'display + trend' selections. The average is 83 per hour and does not vary much over the 14 days of the event recording. The average of the CD-unit is higher than the average of the HV-units (resp. 47 and 36). This point will be discussed in more detail later.

The observation data show that the operator spends 29 % of his time at the CD operating console, 12 % at the HV console and 4 % at the supervisory console. This last percentage equals an average time period of 20 minutes per shift. During the interviews the operators said they felt that they had insufficient time to sit down behind the supervisory console. They said that they collected the long-term supervision information from the logging typewriters and the short-term supervision information from the alarm supervision panels and the CRTs on the operating consoles.

Most operators stressed the point that they missed the parallel (conventional) information presentation, even after they had been working with ICOSS for one year. To update themselves about the process they considered repeated variable selection a necessity. They highly appreciated the presentation of the trends of three related variables together with the information of the variable selected.

As already mentioned, the operator spends much more time at the CD console than at the HV console. The alarm and deviation message data indicate that the HV-units were more stable during the observation period than the CD-unit. In addition, the interview results suggest that the operators pay more attention to the CD because it is a primary unit. This smaller amount of time spent at the HV console is accompanied by some differences in the use of the console facilities. The relative frequency of use of the matrix keyboard is three times that of the CD console. The same applies to the frequency of use of the message VDU on the HV console. These results indicate that when a deviation message or alarm has called his attention, the operator starts working at the HV console more often than at the CD console. When there is a deviation message and the operator presses the flashing group supervision button, information about the off-target variable is presented on the message VDU. If the operator now wants numeric and trend information, he can use the target number presented already on the VDU to select the information via the matrix keyboard instead of using the graphic selection panel.

This different way of working at the HV console is also shown in the way he uses the pictorial displays. The pictorials were originally implemented as a diagnosing aid to provide parallel information. Table 4 shows how pictorials are used by the operator at the two operating consoles.

At the HV console a pictorial is inspected only once in 48 % of the cases. At the CD console this happens in only 11 % of the cases, so at the HV console repeated inspection of pictorials is far below the number of times this happens at the CD console. This means that pictorials at the HV console are mainly used as a diagnosing tool, whereas at the CD console, they are more often used to keep track of changes in a group of measurements. When the pro-

TABLE 4 - Observation Results of the single and repeated Use of Pictorial Displays at the CD and HV Consoles.

| Number of Observations per Pictorial | CDU | | | HVUs | | |
|---|---|---|---|---|---|---|
| | Number of Pictorial Selections | Total Number of Inspections | % | Number of Pictorial Selections | Total Number of Inspections | % |
| 1 | 101 | 101 | 11.3 | 106 | 106 | 48.6 |
| 2 | 64 | 128 | 14.3 | 25 | 50 | 22.7 |
| 3 | 36 | 108 | 12.1 | 6 | 18 | 8.2 |
| 4 | 22 | 88 | 9.9 | 5 | 20 | 9.1 |
| 5 | 16 | 80 | 9.0 | 3 | 15 | 6.8 |
| 6 | 10 | 60 | 6.7 | - | - | - |
| 7 | 3 | 21 | 2.4 | - | - | - |
| 8 | 3 | 24 | 2.7 | - | - | - |
| 9 | 1 | 9 | 1.0 | - | - | - |
| >10 | - | 280 | 30.6 | 1 | 10 | 4.5 |

cess conditions make it necessary for the operator to pay much attention to one of the processes for which he is responsible, his behaviour shows signs of the management by exception approach. He is inclined to attend to the less demanding process only when it calls his attention.

In general the evaluation results do not support the assumptions of the system designers about the work procedure of the operator. Operators have complaints about the control room. The supervision console is of little help. The alarm supervision panels are badly positioned; these panels are high above the operating consoles and at an angle for easy scanning from behind the supervision console. (After the evaluation these panels were replaced with smaller versions in a lower position.) The change from parallel to serial information presentation is largely responsible for the frequent request of point information. This is the less efficient equivalent of walking along the conventional panel.

(2) The effect of equipment alarms and deviation messages.

The alarm rate during the observation was high and did not change very much with time. The average was 13 alarms per hour, with extremes of over 60 per hour. Dangerous process conditions can develop during periods with oscillating alarms, because operators tend to acknowledge alarms prior to determining their location on the alarm panel. If alarms occurring within 30 seconds after each other are excluded, the average drops to 4 alarms per hour. The high alarm level made it difficult to relate the occurrence of an alarm to changes in the operator's behaviour. It was impossible to distinguish the effect of a particular alarm from the effects of previous alarms, because they happened so close together.

The target deviation messages also have a very high rate. The average is 35 messages per hour, 14 per hour for the HV-units and 21 for the CD. It is impossible for the operator to react to each individual message. The operator ignores the messages on the message typewriter. He looks at the message typewriter output within 60 seconds after a deviation message has been printed only 25 times out of 291. Use of the group supervision buttons to request the same information on the message CRTs has, according to the event counting, an average frequency of 6 at the CD console and 7 at the HV console. The interviews, however, indicate that the operator is not totally unaffected by the deviation messages. It appears that he waits until a number of messages have been received from a section. At a certain moment he decides to act. This moment depends to a large extent on the current dynamic state of the process.

In the opinion of the operators the great number of deviation messages and equipment alarms interrupt their personal task scheduling and long-term work planning.

Conclusions relating to ICOSS. The work procedure followed by the operators differs on many points from the one expected by the system designers. Prominent among these points is that the need felt by the operators to have information presented in parallel is greater than expected. The high equipment-in-danger alarm rate during the observation period made an analysis of the operator's behaviour as a reaction to an alarm impossible. The individual deviation messages are ignored, mainly because of their very high number. Only repeated messages from a section of the plant in combination with the current state of the process can induce the operators to take action.

## GENERAL CONCLUSIONS

Operators do not like to have their display facilities organized in such a way that it only allows them to zoom-in on a problem. They have a definite need for information presented in parallel to update their knowledge of the dynamic status of the process. Deviation messages and alarms are of little help here, especially if there are so many that it lowers the credibility of the alarm system. Basically, the operator considers it his task to prevent alarms, whatever the kind. This is the reason why it is not enough for him only to be told what is wrong. He wants to be able to verify that nothing is happening which might eventually result in an alarm or disturbance. The model of operator behaviour currently used by instrument manufacturers in the design of VDU-based control panels is not confirmed by the results of the present study. Although manufacturers developed their model from observations in conventional control rooms, some with VDUs, it seems that the use of this model for fully VDU-based control systems is not appropriate.

Accepting computerized systems with their obvious advantages, does not imply the acceptance of the principle of successive information presentation. There is a need for presentation in parallel and solutions have to be found to satisfy this need. Current systems can present information to assist the operator once he knows the general location of a disturbance. Ways will have to be found to help him to update his knowledge of the dynamic status of the process in a more efficient way than he does now.

## REFERENCES


1. Fraser, G.L., 1978 "A New Approach to the Operator and Control system Interface", Proceedings of The Operator-Instrument Interface Symposium of The Institute of Measurement and Control, Teesside Section, Middlesbrough.

2. Reynolds, B., 1978 "Operator interfaces for centralised control rooms", Proceedings of The Operator-Instrument Interface Symposium of The Institute of Measurement and Control, Teesside Section, Middlesbrough

3. Meer, A.T. van der, 1975 "Computer services Shell's plant operation", Vth Convencion Nacional de la Industria Quimica, Barcelona.

4. Bos, J.F., Jansen, J.P. and Maarleveld, A, 1977 "ICOSS: An Integrated Approach to Process Automation", Proceedings 5th IFAC/IFIP Conference.

Figure 1a  A PO control panel

Figure 1b  The PRIMA interface in the control panel

Fig. 2 ICOSS control room - General view and supervision console

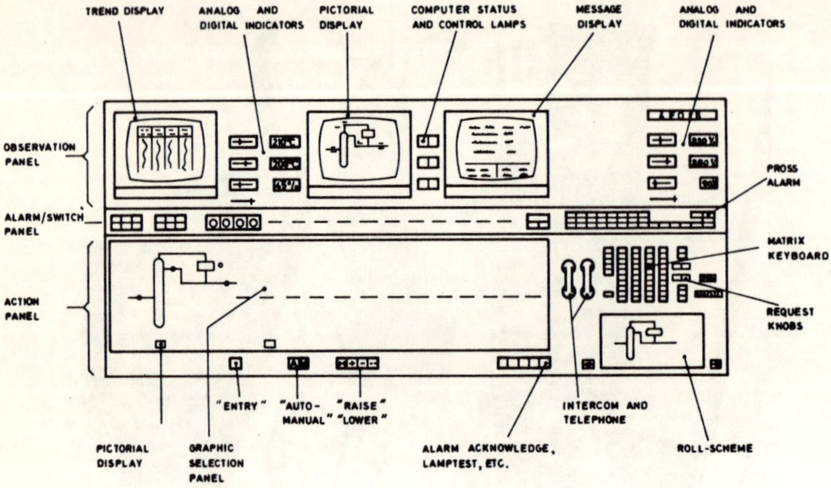


Figure 3  Operating console lay-out

WORKSHOP: EXPERT SYSTEMS IN PROCESS CONTROL

The current interest in Expert Systems has stimulated far more publications than real, live, large-scale systems! This appears particularly true in the area of process control. There is also an understandable reluctance of those organisations that do have working systems to rush into print. The aim of the workshop is to stimulate discussion of some of the many issues raised by the application of Expert Systems.

## INTRODUCTION

(Convenor: P. K. Andow, Department of Chemical Engineering, Loughborough University of Technology).

Expert Systems appear to be applicable to 3 areas of process control:

1. Fault Diagnosis - see paper by Andow elsewhere in these proceedings. Fault Diagnosis (or "Alarm Analysis" or "Disturbance Analysis") might be on-line or off-line.

2. Providing operating advice when changes occur in plant inputs (e.g. feed quality, utilities) or changes are required in plant outputs (e.g. different product specification or alternate product - perhaps on a batch plant).

3. Providing operating advice not directly related to inputs or outputs (e.g. corrosion monitoring, condition monitoring of mechanical equipment, recuperator change scheduling, catalyst activity monitoring and replacement etc.)

The workshop will briefly review these areas in order to set the scene for discussion of particular aspects of Expert Systems. It should be noted that none of the activities above is "new". Decisions on catalyst replacement, for instance, have to be taken on many plants and somebody has to make the decision. If he or she is making "good" decisions - however that might be judged - then a captive expert is available. If no expert is available (or willing to help) then it is not possible to implement an Expert System! This may help to identify the need for particular staff skills or to at least focus attention on the quality of decisions that are being taken.

## SUPPORTING DECISION MAKING AND PROBLEM SOLVING DURING RARE HIGH RISK EVENTS

(Contributors: D. E. Embrey, Human Reliability Associates Ltd, and P. C. Humphreys, Decision Analysis Unit, London School of Economics).

The difficulties encountered by an operator in handling rare, high risk process plant abnormalities will be discussed. It is proposed that in these situations attention has to be focussed on the critical decisions that have to be made and how these decision can best be supported in order to manage the situation and bring the system to a safe, stable state. During rare events the operator is faced with the problem that formal procedures will probably not exist for handling the situation, and if this is outside the operator's experience base he will probably not be able to utilise learnt heuristics and rules of thumb in order to manage it. In addition, as the incident progresses, the evidence available to the decision maker may become increasingly degraded as sensors become inoperable or erratic.

Another problem is that there is unlikely to be any direct indication of the cause of the abnormality. For example, the loss of inventory due to a pipe break, will have to be inferred from the changes in state of a number of indicators. If the event cannot be mitigated by the operating crew, the focus of decision making activity may move from the individual operator to other groups of individuals (e.g. designers, plant managers, safety regulators, local authorities etc). The representation of the problem is likely to be very different for these categories of decision makers and this needs to be taken into account when decision support systems are being designed.

Recently developed approaches for providing a decision support system appropriate for rare events will be described. These start from the expertise possessed by experienced plant operators, which is elicited by means of an interactive computer system to produce an explicit model of their knowledge base regarding the relationships between indicators, causes and consequences. This can be used to model the diagnostic process where the operator starts from an observed pattern of indicators and attempts to infer events which may have led to this pattern.

This process involves the operator exploring his knowledge of consequence to cause linkages, where he looks for common causes of events which may have given rise to the observed pattern of indicators. This is necessary in order to predict future plant states which could result from actions which he might want to consider in attempting to handle the abnormal events.

A computer programme called LINK is used interactively with appropriate experts (e.g. operators) to elicit the cause-consequence-indicator data base. Another programme called EXPLORE can be used to explore the network generated by LINK. To use EXPLORE, the indicators which are present, and those which are definitely absent, are input to the system. This then examines possible causal sequences (including other consequences from the same common cause) which could account for the observed pattern of indicators. The system generates a comprehensive report as it evaluates possible cause consequence sequences. The aim of the system is not to generate an unambiguous diagnosis of the cause of the situation, but to stimulate appropriate 'what if' questions. These in turn may suggest additional types of information which the operator may need to acquire in order to choose between alternative courses of action to handle the event.

Another programme called CASCADE uses Bayesian techniques to calculate the probabilities of the various causal paths identified by EXPLORE, if the prior probabilities can be assigned to links in these paths. CASCADE can also warn if the operator model captured by LINK is inappropriate for the situation being evaluated.

In addition to their use in supporting operator decision making, these approaches also have applications in training, the development of procedures, and design of the man machine interface. These will be discussed in the workshop.

## FURTHER ISSUES FOR DISCUSSION

(Contributor: P. K. Andow, Department of Chemical Engineering, Loughborough University of Technology).

Expert System applications are rare, particularly in process control. The technology and its spin-off problems are relatively unknown and technically immature. Many issues appear ripe for discussion in a workshop session. Even successful applications are often not reported in the literature for reasons of commercial confidentiality. How many failed applications are there? Why did they fail? The aim of this final section of the workshop is to raise some of these issues and to seek opinions on them - solutions will take a little longer!

As a starting-point the following topics are suggested:

1. Domain knowledge:

   a) How well-defined are the "rules" for the application areas noted earlier?

   b) Do we have good heuristics available where large problem-spaces exist?

   c) Is the knowledge properly-tested and "complete"?

2. Software:

   a) What kind of system should we use? (Production rules, logic programs, semantic nets, frames).

   b) Is there a "shell" available that suits the application in hand? Do we "start from scratch"?

   c) Is the problem size manageable with the available software?

   d) Can the system be easily and thoroughly tested after changes are made?

   e) Will the response-time be adequate?

3. Implementation Issues:

   a) What kind of dialogue is needed so that the system will work properly?

   b) What are the implications for the operator's task?

c) What training is needed and what sort of "operator" is the system aimed at? (control room operator, shift supervisor, plant engineer?)

d) How well can the system be integrated into the whole monitoring and control activity?

e) Can the system be designed so that it enables the operator to make better decisions in a way that he perceives to be complementary to his own skills?

There will be many other issues that can affect the effectiveness of a system. The workshop is intended to provoke a free and wide-ranging discussion.

WORKSHOP: COMPUTER-BASED SYSTEMS IN PROCESS CONTROL

This workshop reviews ergonomics implications of computer-based display and control systems. Several questions on operators' interactions with such systems need to be asked, particularly about the display reading and monitoring, and response to alarm information. For one particular system, the introduction of VDUs has been successful, though some problems in workspace design and in alarm presentation have been noted. Another study has emphasized the role of console design in promoting efficient performance of both individual operators and groups of operators. Finally, there is emphasis on the needs to take account of commercial influences, and to provide adequate ergonomics guidance for system designers.

## OPERATOR-VDU INTERACTION IN PROCESS SUPERVISION AND CONTROL

T.M.J. Lenior, R.M.C. Paulussen, J.H. Poesse, J.E. Rijnsdorp
Ergonomics Group, Twente University of Technology, Enschede, The Netherlands

### Summary

The change-over from conventional to VDU-based instrumentation modifies the supervision and control task of the control room operator. More insight can be obtained from observations, protocol analysis and interviews. The results can be used for defining laboratory experiments with human subjects.

### Introduction

The widespread introduction of distributed information systems in the process industries has a strong impact on the operator's job. Instead of monitoring many recorders and indicators, he is now confronted with a small number of visual display units (presently usually CRT's). This involves, amongst others, a change-over from easy access to a large area of values displayed in parallel, to more difficult access to a small area of serial displays. Here it is important to evaluate the different ways of putting pictures on the CRT-screens: menu selection, direct commands, stepping through a series of related pictures, etc. In other respects there are also many differences between conventional and VDU-based instrumentation.

More insight can be obtained in these matters by a combination of field work and laboratory experiments. The field work consists of observing the operator's activities, asking him why he performed certain actions, and interviewing him about his experiences and opinions with respect to VDU-based instrumentation. In addition, system designers and application engineers will be interviewed too. In this way, difficulties are pinpointed, and operator interface tasks are described more precisely.

On the base of these actual tasks, experimental tasks are defined for human subjects in laboratory experiments, which can yield quantitative information about important trade-offs between alternative ways of structuring CRT-pictures and operator/VDU-dialogues. Results of these experiments will be published in the near future.

The next section gives more details about the field work being done by the Ergonomics Group of the Twente University of Technology. Results will be presented during the symposium.

## Field Work

In the past, many field studies have been performed in different countries and industries for the case of conventional instrumentation systems (ref. 1). A common finding is the substantial amount of time the control room operator spends in monitoring displays, irrespective of day or night, beginning or end of the shift periods, the degree of automation, etc. Protocol analysis (ref. 2)of his verbalized thoughts has shown a correlation with the question: is the process state all right (ref. 3,4)?

Presumably, during normal operation, the operator tries to detect impending off-normal conditions as soon as possible (ref. 5). He does not wait for alarm signals, probably because this leaves less time for decision-making, and only occasionally offers really urgent information (ref. 6).

Another problem is the real meaning of 'process state' for the operator. His preference for recorders over indicators suggests estimation of rates of change of process variables, in addition to present values (ref. 7). In general it can be said that the control room instrumentation is mainly used as an information system, and less often as a manual control system.

## Questions to be answered by Field Work

For the field study of VDU-based process instrumentation the following questions are important:

a) How do operators select and scan CRT-pictures? Do they follow certain patterns, or is there a strong correlation with experienced or envisaged difficulties?
b) Do operators utilise overview pictures? If so, do they do this mainly as an alarm overview, a present-value overview (actual value of deviation from set values), or as a dynamic presentation (observing jumps in displayed values at the moments of display updating)?
c) How do operators respond to alarm information? Do they consider this information as helpful, confusing, or, in some cases, as a means for communication with field operators (see ref. 5 for an example).
d) How often do operators monitor trend recorders or (conventional) recorders? Do they look for gradual drifts, sudden changes (caused by malfunctioning process equipment or instruments), and for variation patterns?
e) How do operators name process variables: by tag number or by process equipment (e.g. 'reflux flow of column 706')? In how far do they really use 'names' during interaction with the interface (spatial coding could provide the required information).

f) If present, how do operators utilise process computer displays? Are these more, or less important than the process instrumentation displays?

Note that the first question of every item can be answered by observing operator behaviour only, while dealing with the second questions requires 'thinking aloud' or interviewing.

In general it makes sense to design operator/computer/process interfaces for critical situations, viz. large upsets in process operation. Unfortunately (from the point of view of research only!) the probability of observing an upset situation is rather low, hence information must mainly come from interviewing the operator, and staff people with a good knowledge of day-to-day process supervision and control.

Preliminary Conclusions

- Assessing the process state is a central operator task.
- Very probably this encompasses present values and rates of change of process variables.
- Observation of process operator activities can be supplemented by protocol analysis of his verbalised thoughts and/or by interviews.

References


1. See: E. Edwards, F. P. Lees, The human operator in process control, Taylor & Francis, 1974.

2. L. Bainbridge, Analysis of verbal protocol from a process control task, in ref. 1.

3. H. Kragt, J. A. Landeweerd, Mental skills in process control, in ref. 1.

4. H. van Droffelaar, A field study of stress, experienced by operators supervising a highly automated process, in: Schuh and Sprague (Eds.), Productivity and Man; IFAC workshop (RKW, Frankfurt, 1975).

5. V. De Keyser, La fiabilite humaine dans les processus continus, les centrales thermo-electriques et nucleaires, CERI, Brussel, 1981.

6. H. Kragt, Operator tasks and annunciator systems; PhD thesis, Eindhoven, 1983.

7. J. E. Rijnsdorp, Structuring process information on VDU's; Workshop during Human Factors Society 26th Annual Meeting (Seattle, 1982).


# INVESTIGATION INTO THE ERGONOMIC PROS AND CONS OF A VDU-INTERFACED PROCESS CONTROL SYSTEM BY MEANS OF OPERATOR INTERVIEWS: A CASE STUDY

E. Piso
DSM-Research and Patents, Geleen, The Netherlands

Summary

This study can be divided into two parts: (a) console evaluation and design; (b) evaluation of the system. Part (a) was conducted by means of group interviews and user participation and resulted in a large VDU-console design. Part (b) was investigated in individual interviews with users. This resulted in several pro's and con's of the system. Especially the presentation of information and alarm messages was criticised.

## 1. Introduction

In this case study a VDU-interfaced Process Control System (PCS) was applied on a rather small scale. The process can be divided into six more or less independent parts.

Only one, the reaction part, was instrumented with a VDU station. The other process parts were still instrumented in the conventional way, with pneumatic controllers placed on wall panels together with indicators, paper recorders and annunciator fascias. The lay-out of control room and VDU console is shown in fig. 1. After it had been used for two years, the system had to be evaluated ergonomically, for two reasons. Firstly, there were plans to instrument the other process parts also with VDU's within the next five years. The console had consequently to be extended with several VDU stations in an ergonomically justified way. Secondly, management and staff were interested in the pros and cons of the system in view of the envisaged increase in its application in several plants.

## 2. Method

This case study can be divided into two parts, dealing with: (a) the console evaluation and the design of a large console; (b) the evaluation of the PCS system, its use, and the pros and cons inherent in the system or accidental to it.

### To a. Console

It was decided to ask the operators employed on four shifts in a plant for their opinions on the VDU console. In group interviews held with each shift such console aspects as shape, location in the control room, place of the VDU and other instruments in the console, writing surfaces, lighting, problems arising when several people use one console simultaneously, and still other aspects were evaluated. Afterwards the groups of operators were asked to design a large console using two-dimensional cardboard console-units (scale 1:10). Also, a preferred location on the control room map was to be indicated. The interviewer helped the operators by presenting to them the pros and cons of the generated alternatives.

The drawings made by the four groups were not essentially different. In so-called feedback-interviews the groups reached agreement on one final design.

### To b. System Evaluation

In order to evaluate the system properly, individuals (not group) interviews were held with all types of users. Three operators per crew/shift were asked to co-operate. They all had experience with the system. The individual interviews lasted about two hours each. The main topics were:

a) introduction and instruction;
b) presentation of information;
c) alarm presentation;
d) keyboard;
e) summary.

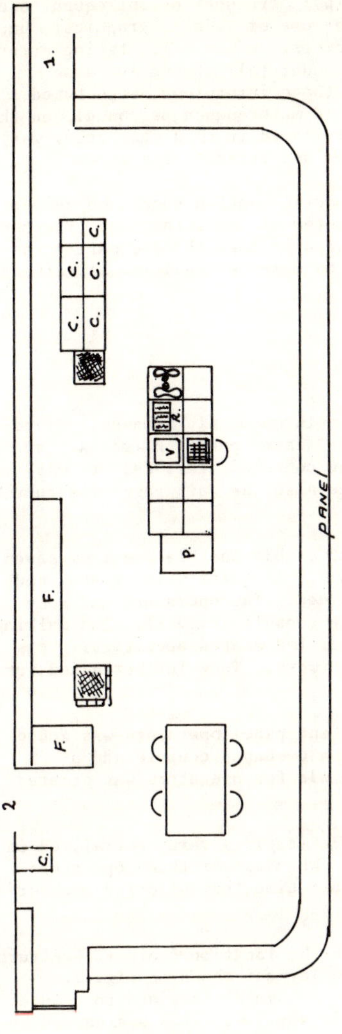


FIGURE 1: Old control room lay-out

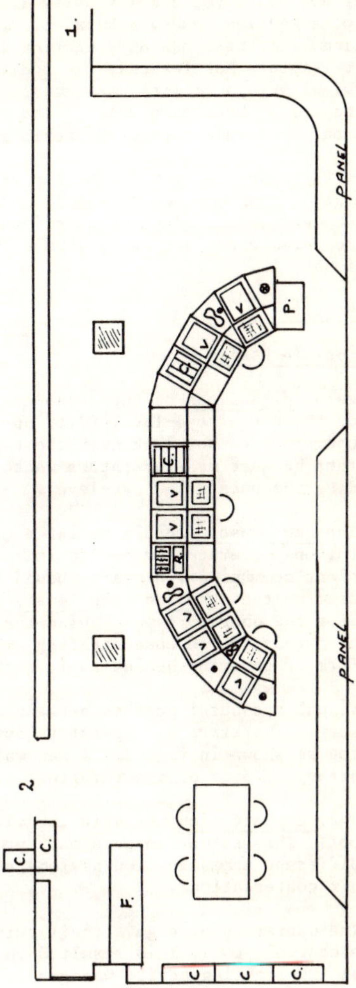


FIGURE 2: New control room lay-out

Legenda: V = VDU-station
P = Printer
F = Flow-schemes
C = Cup-board
R = Recorders
1 = Entrance factory
2 = Entrance cantine
⊠ = Microphone

Items dealt with under topic (a) were, for instance: instruction manuals, training schemes, user-participation; under (b): frequent or infrequent use of certain displays, display appreciation, the use of trends, graphics; under (c): presentation modes, acknowledging procedures; under (d): typing errors, function keys, response and reaction times; under (e): strong and weak aspects, future bottlenecks. In addition to these interviews, structured interviews were held with management, staff and maintenance personnel, on the same topics as mentioned above. The method followed in this case study was based on experience gained in previous studies and literature sources.

With regard to part (a) of our study, special mention should be made of the control room project (Piso, 1) and some internal investigations (Bouwman, 2). Concepts underlying and questionnaires used in part (b) are partly based on a comparable evaluation by Thomas (3) and on some internal investigations (Cellissen, 4).

## 3. Results

### 3a. Console Design

The design dimensions of the console satisfy anthropometric demands. Placement of the VDU at eye-level (with operator in seated position) means that many operators cannot look over the top of the console. This feature was important because many operators wanted to see also the wall panel instruments, including the ones below eye-level.

The new console should be large enough to enable the placement of seven VDU stations, operated by one to four operators. Adequate writing area and sufficient communication means should be provided. The operators had some criticism on the shape and the location of the console (fig. 1). The walking aisle for the operators was behind the backs of the seated operators. This fact hindered them in concentrating on the VDU work. They further asked for some form of covering behind their backs.

Visual and aural contact between console and panel operators was found necessary. Therefore the operators opted for a C-shaped console and a location as shown in fig. 2. A new walking aisle for operators was created by rearranging part of the furniture.

The communication means in the old console (fig. 1) were centralized in one spot. This situation gave many problems. When two or three operators used different communication systems at the same time they hindered each other in their conversations.

The operators also gave their opinions on the locationof other furniture in the control room. This resulted in the new lay-out shown in fig. 2. Special VDU-lighting advice was given in order to avoid glare and to reduce straylight problems. The contrast on the VDU's was low. This was caused by the high level of illumination in the control room, which, however, was needed for writing and panel work.

### 3b. System evaluation

#### Introduction of the system

The operators experienced the small-scale implementation of the VDU system as very favourable. Hence, the change to a completely new control system was gradual and progressive. After a theoretical instruction, the system was learned on the job. Young operators had the impression that they learned the system faster than the older operators did.

However, all ages proved to be able to learn it well. There seemed to be a need for a 'second' theoretical instruction after a few years, to get to know all the ins and outs of the system. Many operators found working with such an advanced system very interesting. But this effect was only temporary.

## Presentation of information

Like many other systems, this system has a hierarchy in its displays. In the overview 256 control-loops can be displayed by bar-graphs. A group display consists of eight loops (or indicators), presented by bar-graphs and digital information in engineering units. The upper part of this group display shows eight squares (a so-called bonus display) which can be used for indicators or digital contacts. Displays on the 'loop' or 'input' level will not be dealt with here. There are also trend displays and graphic displays, as well as an alarm summary display.

The operators used the group-display only. They regularly displayed the six groups on the VDU in a row. The most important display, that for the final reactor phase, was left on the screen for monitoring. Because of the limited number of control loops and indicators, it was not necessary to use the overview display. The bars in the overview display were considered to be too close to each other. Variables were therefore difficult to identify.

The trend displays were used for tracing back disturbances. However, because of the limited screen size and memory capacity, process values are mathematically averaged. Hence, brief disturbances cannot be seen on the three-hour or the thirty-hour trend display. The display of three variables in one diagram was very confusing in the identification of variables and determination of the process values. Three paper recorders connected to the system were used only for some important variables.

The grphic displays (flow schemes) were not used. Although they contained live-updated process values, they were not found to be interesting in the plant here under consideration (but they may be with other processes). Careful advance consideration of the pros and cons of graphics for individual plants may lower software cost.

## Alarm presentation

The identification of alrams was rather difficult. Many digital alarms were presented by means of internal system codes. With reference to a list the operator had to translate this code into a tag number and a description. Some software changes in the system made it possible to use the bonus display for presenting alarms of digital contacts. A group display now shows great resemblance to a conventional wall panel: eight loops or indicators and a small fascia with eight lamps. Acknowledging of alarms should only be possible when the display with the on- or off-going alarm is on the screen. Central acknowledging by simply pressing a button is dangerous.

## Keyboard

The keyboard is of the membrane type. All switches are covered with a pad. It was chosen because of its resistance to the effects of dirt, dust and coffee. This keyboard was judged to be good for the purposes operators use it for, e.g. controlling loops and asking for information. It is not suitable for software configuration.

If the pad, or membrane, is completely flat exact finger positioning poses a (minor) problem, which can be removed by providing a small cavity at each key.

## Summary

The most positive aspects of the system were the speed and accuracy of the control actions, especially compared to the pneumatic system. A negative aspect was the alarm identification.

Everybody concerned with our study had confidence in the future of this control room in which all wall panels will be replaced by the VDU systems.

## References

1. Piso, E., Ergonomic evaluation of control rooms: two case studies. In: IFAC 8th Triennial World Congress, Kyoto, Japan, Edited by H. Akashi, Oxford, Pergamon Press (1983), pp. 3455-3461.

2. Bouwman, B., private communication (1982)

3. Thomas, D. B., An assessment of the adequacy of the CRT as an information display medium for process control; (1982) Koln, Tuv, EEC report VIS/81-502, nr. 5908/82 E.

4. Cellissen, F., private communication (1982).

# THE IMPACT OF PROCESS INTEGRATION AND COMPUTERISED INSTRUMENTATION ON OPERATOR CONSOLE DESIGN

T. M. J. Lenior, R. N. Pikaar, J. E. Rijnsdorp
Ergonomics Group, Twente University of Technology, Enschede, The Netherlands

## Summary

Process integration imposes stringent requirements on control room design, and console lay-out is also influenced by operator tasks, and by conditions during large plant upsets or emergencies. In view of ergonomic advice in several control room design projects, commercial CRT-based instrumentation systems were investigated. Some of these do not meet all relevant ergonomic requirements for process supervision and control.

## Control Room Lay-Out

Particularly in large-scale processing plant, energy and materials can be saved through process integration. For instance, waste heat in one process unit can be transferred to another one; waste gas generated in one location can be combusted at other locations, etc. However, the resulting interactions make the total plant more vulnerable to upsets, and fault management more complex. Hence good co-operation between control room operators is essential.

In recent control room design projects, these considerations have led to the choice of certain solutions for the balancing of technical and social/organizational requirements.

Mostly, the operator work load determines the number of operators in new control rooms. A recent field study, related to one of the projects and performed in the existing control room, has indicated that the operator work load can be estimated from the number of automatic control valves[1], the number of associated field operators and the number of important process streams (for more details, see ref. 1). Moreoever, peak load can be reduced by providing means for mutual assistance. This has led to 5 consoles, each not only suitable for one operator when conditions are steady, but also for two operators and their supervisor during large process upsets. The distribution of process units over these consoles, and their relative positions, have been determined on the basis of the numbers of interconnecting process streams, and the proposed work organisation.

For smooth co-operation between operators, direct visual and aural contact was considered to be necessary. With so many consoles, this is possible if they are placed not too far apart and if they can look from a sitting position over the top of each console in several different directions, ie. to all other operators.

Vertical Console Lay-Out

Modern process automation encompasses two levels:
- distributed instrumentation systems which replace the traditional controllers, indicators, and, up to a point, recorders;
- computer information and control systems, which perform the higher level functions.

For communication with the console operators, both levels are connected to CRT's and keyboards. Of course, a number of conventional recorders, annunciators, switches, telephones, etc. have to be accommodated too.

The vertical viewing angle for characters should be chosen well above the minimum (15 arc min according to Stewart, ref. 2) as several CRT's have to be monitored by different people, sometimes simultaneously. Hence, with a desirable viewing angle of about 20 arc minutes for characters, and 24 lines of text on the screen (50% spacing), the CRT requires a total perpendicular viewing angle of about 15 degrees (from top to bottom, see Fig. 1). When the CRT's are arranged in a row on the top of a desk, this fits in rather well with the ergonomically desirable viewing direction: 15-25 degrees downwards[2]. The surplus angle is smallest in the case of short persons, who should still be able to have visual contact with operators at other consoles. Hence, the top rim of the CRT-case must be very thin, preferably not more than 3 degrees of viewing angle in height.

A good sitting posture is with upper parts of the legs and lower parts of the arms in horizontal position. For the 95-percentile of Dutch males, 30 years of age in 1986, this requires a desk top height of 0.74 m, and a vertical clearance below the desk of 0.70 m for leg space. However, the 5-percentile of Dutch males, 60 years of age in 1986 would need a desk top

---

[1]In this context, automatic control valves (or other types of correcting units) are preferred over control loops because the latter concept loses its precise meaning when high-level control is implemented.

[2]There is much variety in recommended angles (see, e.g. ref.2). We have followed the VDU-guidelines of the Dutch Ergonomics Society (ref. 3)

height of 0.69 m. This constraint can be met by utilising an adjustable chair and a foot rest (see Fig. 1). In this way, a 5-percentile person has his eyes at 1.19 m above the floor, hence 0.45 m above the desk top.

Another consideration is the amount of work space in front of the operator. With 13 inch screens and 5 mm character height, the viewing distance, corresponding to 20 arc minutes (see above), becomes 0.860 m. Accounting for oblique viewing, this leads to a viewing distance of 0.800 m for the 95-percentile person. Projected horizontally and subtracting a not too unreasonable stomach space for the 5-percentile person, yields a desk depth of not more than 0.55 m. With 0.35 m required for reading and writing (strictly minimal for A4 paper size!), only 0.2 m is left for the (movable) keyboard. Taking a keyboard slope of 16 degrees (US Mil Std 1472 B), this leads to a keyboard depth of only 0.21 m, which is much less than found in most commercial systems.

If one tries to solve this problem by increasing the size of the screen, there appear new problems with keyboard manipulability, and screen positioning on the desk. Moreover, as the operator has to work with several CRT's (see next section), a large distance between keyboard and screen causes large parallaxes, hence it restricts interaction possibilities where spatial D/C-compatibility is utilized.

## Horizontal Console Lay-Out

The horizontal lay-out is strongly influenced by the tasks the operators have to perform. A central activity is assessing the process state (see ref. 4 for further references): even during normal operation, the operator watches present and past values of process variables in order to detect impending off-normal conditions as soon as possible.

Conventionally, this information is provided by rows of recorders, arranged for quick scanning. Modern instrumentation systems offer trend displays on CRT-pictures, but require much button-pushing in order to overview the hundreds of important process variables allocated to one console operator. Evidently, parsimonious and user-friendly screen utilisation and efficient keyboard commands are very desirable here (ref. 4).

Off-normal conditions can be roughly divided into two types:

- more or less gradual deviations from targets for product quality and process efficiency;
- effects of faults in process equipment, instrumentation, computers, and the like. Also large upsets, such as those caused by power failures, fall in this category.

Coping with both types of off-normal conditions can require looking at interactions between process variables, which can best be observed in CRT-pictures dedicated to specific process units (such as a distillation column, a chemical reactor, etc.). Often there is trouble with one or two of such process units, hence having two additional CRT's is not a luxury. Together with a CRT for assessing the process state and alarms, and another one for high-level control, this leads to a minimum of four CRT's per console.

For effective viewing of four CRT's from one sitting position, the displayed characters should be wide enough for readability at an angle. Partial compensation can be found in putting the CRT's in a slight arc, but the console should still be suitable for at least two sitting operators, together with their supervisor in standing position (who is called in when

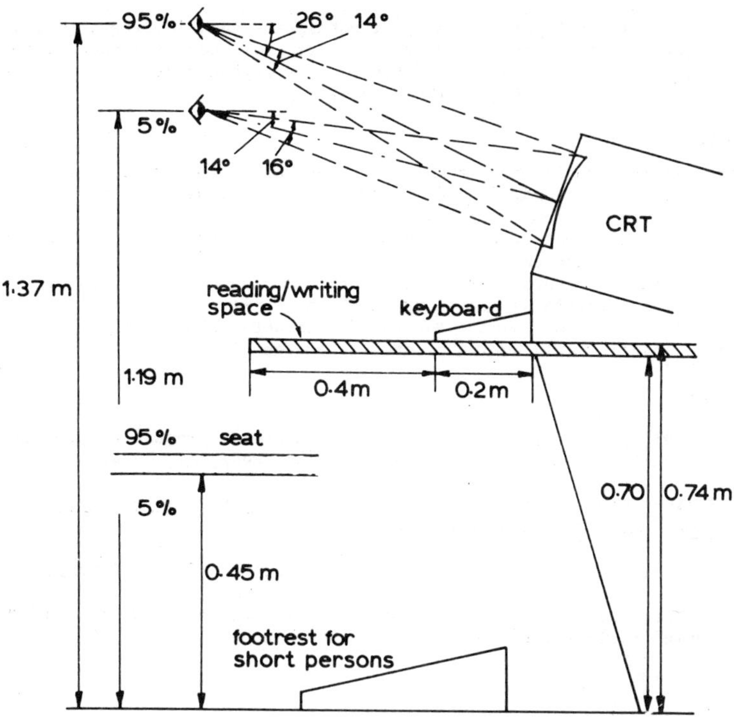


Fig. 1: Sketch of vertical console layout, suitable for 5% (60 yrs.) and 95% (30 yrs.) of Dutch male persons (1986).

conditions become very complicated or critical). In this situation, narrow characters (with a width less than 70% of the height) are very difficult to be read by these persons simultaneously, hence more or less square characters are required. Unfortunately, many commercial CRT-systems do not meet this requirement.

With four 13 inch screens in tight cases, the console length is 1.5 m, not counting additional space for annunciators, switches, telephones, intercoms, etc. Of course, a length of about 2 m is anyhow required for two seated persons working side by side at the console.

## Conclusions

- Process integration requires further centralisation of process supervision and control, and direct aural and visual communication between control room operators;
- Direct visual contact between many console operators is possible when CRT's are tightly encased.
- Many commercial keyboards are too large in view of leaving sufficient reading and writing space in front of the operator.
- Many commercial CRT's have too narrow character shapes for viewing at an angle (particularly during upsets and emergencies).

## References

1. J. E. Rijnsdorp, R. N. Pikaar, T. M. J. Lenior; Process Control and Organization; Submitted for IFAC Congress, Budapest, 1984.

2. Ericsson Information Systems; Ergonomic principles in office automation; Bromma, Sweden, 1983.

3. BEA list; Dutch Ergonomics Society (Amsterdam, 1979, in Dutch).

4. Rijnsdorp, J. E.; Structuring process information on VDU's; Workshop at Human Factors Society 26th Annual Conference (Seattle, 1982).

# THE IMPACT OF ERGONOMICS RESEARCH ON THE DESIGN OF THE OPERATOR INTERFACE

M. S. Carey
Ergonomics Development Unit, University of Aston in Birmingham

## Summary

The main contention of this contribution is that the widespread adoption of an ergonomics principle into the design of operator interfaces relies on more than the essential ergonomics justification. Two issues are discussed in relation to this. Firstly, the weight given to ergonomics factors by the manufacturers of process systems depends directly on the interests of their customers. It is suggested that providing both the purchasers and the vendors of process control systems with techniques for evaluating the quality of the operator interface might have beneficial effects in terms of the status and practice of ergonomics during design. Secondly, the ease of application of research findings to specific design questions needs careful consideration. Recent trends towards generalisable theories and techniques for use in interface design are described in terms of their roles as communication channels between the researcher in ergonomics and the interface designer.

## Introduction

The implicit aim of all ergonomics research is to ensure that some object or system is effective and comfortable to use, through the study of human characteristics. With regard to the ergonomics aspects of the operator interface in process control systems, those in research can be said to have failed to achieve this aim if the knowledge they produce is not eventually reflected in the design of commercial systems. Unfortunately, the brilliance of a piece of research does not on its own guarantee the adoption of its results into working practice. Two particular factors can be seen to play a part in the uptake of ergonomics principles by system designers:

(i) Commercial forces determine whether there is any need or advantage in expending great efforts on the ergonomic aspects of systems.

(ii) The form in which research findings are delivered to system designers may severely limit their scope and usefulness.

The following sections examine these two points in more detail. The information given on the marketing of control systems was obtained during visits to six US control system manufacturing companies during December 1983. The willing co-operation of these companies and their personnel is gratefully acknowledged.

## Ergonomics Issues in their Commercial Context

The low volume and specialist nature of process control applications generally tends to mean that control system manufacturers must tie themselves very closely to the perceived needs and priorities of their marketplace. In a number of companies product specification is directly a role of marketing. The features provided within the operator interface to a process control system, therefore, are strongly dependent on the requirements and preferences of a vendor's current and potential customers.

Within customer organisations, purchasing decisions are often the responsibility of company control engineers or software specialists. These personnel are required to assess both the technical performance of vendor systems and the quality of the operator interface provided. In the achievement of the latter, they are required to make decisions about the suitability of a user interface and display console on the basis of their perception of the needs and abilities of the intended users of the system. Obviously the purchaser's understanding of operator needs will vary from case to case, but in general a number of misconceptions often conspire in the purchaser's mind to bias his view of the value of an interface. For instance, in the absence of any means of comparing the more elusive aspect of 'ease of use' between systems, purchasers will often draw comparisons between interfaces on the basis of gross hardware or software features (such as the size of the colour monitor screen or the existence or absence of process mimic facilities). There is often a doubtful relationship between such features and interface quality. In a similar manner, purchasers sometimes fail to distinguish between the different interface requirements of separate application areas, and may, for instance, be concerned about the size of a display colour palette, expecting to see the range of colours available on special purpose colour displays, whereas sixteen colours may be more than adequate for process control. Finally, the computational background of a purchaser may affect his appraisal of how easy the operator will be able to learn how to operate the system.

The purchaser of process control systems is certainly not alone in facing the problems of assessing the quality of a user interface (Carroll & Rosson, 2) and as a result methods of evaluating interface quality are receiving a sizeable amount of attention amongst researchers at present. Increasing the knowledge and awareness of the factors associated with system quality within customer organisations is likely to have two associated effects: firstly, a more accurate balance will be struck in assessing the relative merits or otherwise of the features and attributes of a particular operator interface and secondly, this ability to come to more rational and accurate assessment is likely to have the effect of raising the importance of the interface amongst other factors when making a purchasing decision.

An increased and informed interest by their customers in the facilities provided for operators in control systems is likely to bring a relatively swift response from system manufacturers. The same information and evaluative techniques supplied to user organisations would equally be utilised by the manufacturers of systems in testing and improving the quality of their operator interfaces.

## Meeting Information Needs in Interface Design

There is a constant danger in the study of human and computer interaction that research may rarely progress far beyond the experimental comparison of input devices or command language styles. Such studies may be important for resolving the particular issues they are set up to investigate, but they produce only a limited amount of information that can be generalised across to other tasks, with user groups and forms of interaction that may be encountered at a later date. To answer a particular problem or to decide between alternatives, an interface designer is often forced to abandon the literature and resort to using his judgement and intuition. As technology progresses and applications change, the total database of research findings is always going to lag far behind the needs of those designing human and computer interfaces.

In order to provide information for interface designers that will not date nearly so quickly and will be more widely applicable requires the pursuit of reliable theories and analysis techniques. As Norman (5) points out, such theories and techniques would be required to be easy to use, justifiable and they must be seen to possess face validity. In essence, they would represent a set of design tools, encapsulating the findings of a variety of observational and experimental studies and providing the interface designer with information on which to make much more knowledgeable and accurate decisions. There appear to be two main classes of such theories and techniques; those that focus on the psychological behaviour of the user of a system and those that focus on the structure of the interface.

The most exhaustive approach to a model of user psychology as yet is described by Card, Moran and Newell (1). A distinctive feature of their work is its quantitive nature; this enables alternative forms of interaction to be compared on the basis of time by forming a model of user behaviour and then determining normative values for its individual parameters. Within the text editing domain in which their model was developed, it appears to be capable of making predictions to a degree of accuracy acceptable for design purposes. Their model is far from complete, as it is based on only one parameter of performance, that of time, and at present has no provision for representing such behaviour as the acquisition and development of expertise by a user. However, a growth in interest in theories of user learning, mental structures and errors promise the construction of further prescriptive models. The extension of such models to the operator interface will require some domain specific additions to the state of present knowledge. At the highest level of the model, additions need to be made to account for the ways that operators perceive and store process specific information, their predominant control strategies, and the way they subdivide and organise their activities to meet overall process goals. At the lowest level of the model, the repertoire of basic mental operations needs to be extended to include visual activities such as pattern perception and visual search. Such models should provide the designer with an accurate and concise grasp of user psychology in a form which is readily applicable to the decisions that need to be made during interface design.

The types of theories and techniques that are emerging with regard to interface structure are quite diverse. The finite state machine has been employed in different ways by a variety of people (eg. Jacob, 3) as a means of representing in a graphical form the flow and structure of the complete set of transactions that are possible between a computer system and its users. A number of alternative interface description grammars have been suggested (eg. Reisner, 7; Moran, 4; Payne & Green, 6) that use the discipline of a formal grammatical representation to demonstrate various aspects of order (or lack of it!) within computer command languages. As an example of how theory can unify and consolidate experimental findings, Rosenberg (8) describes a method of analysis that can be applied to sets of command names to uncover potential sources of confusion and errors before they are put into use. Quite a wide range of different aspects of command naming have been studied experimentally and Rosenberg demonstrates how his theory could account for a sizeable proportion of these. Again, as with the models of user psychology mentioned earlier, though their theoretical background is quite complex, the resulting techniques are quite simple in concept and are intended to be easy to apply.

It is envisaged that such theories and techniques would never represent the entire substance of research into the human and computer interface and are unlikely to make redundant specific research into process interface issues. However, they provide a useful means of summarising research and communicating it to systems programmers. By such means, the output of the fast growing community of those involved in research into the human and computer interface may be chanelled in a form that is acceptable within the constraints and commercial pressures of process control interface design.

References

1. Card, S., Moran T. & Newell A. (1983) The Psychology of Human Computer Interaction, (Hillsdale, NJ: Erlbaum Associates)

2. Carroll, J. M. & Rosson, M. B. (1984) Beyond MIPS: Performance is not Quality. BYTE, February 1984, pp. 168-172.

3. Jacob, R. J. K. (1982) Using Formal Specifications in the Design of a Human-Computer Interface. In Human Factors in Computer Systems, Gaithersburg, March 15-17, pp. 315-322.

4. Moran T. (1981) The command language grammar, International Journal of Man-Machine Studies, 15, pp. 3-50.

5. Norman D. A. (1983) Design principles for Human-Computer Interfaces: In Proceedings of CHI '83 Human Factors in Computing Systems, Boston, December 12-15, (New York: ACM) pp. 1-10.

6. Payne S. J. & Green T. R. G. (1983) The User's Perception of the Interaction Language: A two-level Model. In Proceedings of CHI '83 Human Factors in Computing Systems, Boston, December 12-15 (New York: ACM), pp. 202-206.

7. Reisner P. (1982) Further Developments Toward Using Formal Grammar as a Design Tool. In Human Factors of Computer Systems, Gaithersburg, March 15-17, pp. 304-308.

8. Rosenberg J. (1983) A Featural Approach to Command Names, In Proceedings of CHI '83 Human Factors in Computing Systems, Boston, December 12-15, (New York: ACM), pp. 116-119.

# NOT WAVING BUT DROWNING: PROBLEMS OF HUMAN COMMUNICATION IN THE DESIGN OF SAFE SYSTEMS

Linda J. Bellamy*

This paper emphasises the importance of taking task related interpersonal communication into account when designing safe systems. Potential weaknesses in such communications are illustrated using an information processing model. Specific problems are highlighted by giving examples of errors that have preceded major accidents. The implication is that use of a natural language is a highly imperfect means of accurately communicating task related information, particularly in complex communication structures. Such imperfections need to be recognised and eliminated and the communication process sufficiently formalised to minimise error.

## INTRODUCTION

Ergonomics has not progressed very far in extending the concept of communication beyond that of the interaction between man and machine. There is not an area of ergonomic study that can be readily distinguished as principally concerned with task-related communication between people. The design of procedures, for example, has received little attention in this respect. Although general principles can be applied, such as those derived from studies concerned with the presentation of written information (e.g. Wright and Barnard, (1)) procedures must often be modified according to their shortcomings as discovered when they are actually in use, for example maintenance procedures (Losee et al (2)) and emergency procedures in nuclear power plants (Brune and Weinstein,(3)). It is frequently not only the basic design principles of the procedures that are at fault, such as failing to use short simple sentences or having a "check-off" facility, but rather a failure in accuracy, completeness or organisation of content, or even a failure to have or to use a formal procedure at all. Taking a recent example, the Radiochemical Inspectorate (Great Britain: Department of the Environment, (4)) concluded with regard to the leak of radioactive waste from Sellafield that:

> "The procedure for communicating information between managers by way of manuscript entries in log books appears to us inadequate, prone to error and not sufficiently formalised."(p.13)

Similarly, the Health and Safety Executive Report (Great Britain: Health and Safety Executive (5)) suggested it was:

> "because of what appeared to be a failure of communication between shifts."

Failure of interpersonal communication has frequently been a common contributory cause of major accidents. This author has previously highlighted this fact (Bellamy, (6)). When communication errors occur a "mismatch"

* Ergonomics Development Unit, University of Aston in Birmingham.

situation can develop between the actual state of a system and the perceived state of the system, leading to a failure on the part of an operator to act within the limits set for normal operation of that system. Some examples of the communication errors found were: beliefs that communication was not necessary, use of an informal communication system, a formal communication system being allowed to lapse, blocked communication channels, noisy communication systems being used and incomplete or inaccurate encoding of information. However, it is not the purpose here to discuss accidents in detail although specific examples of communication error will be given. Rather, the intention is to elucidate how interpersonal communication systems operate and to discuss some of the factors which affect communication reliability. The overall aim is to highlight the need for improved systems of task related information transmission and exchange between people.

## INTERPERSONAL COMMUNICATION AND THE INFORMATION PROCESSING MODEL

Shannon's problem, finding solution in the mathematical development of information (or communication) theory (Shannon & Weaver, (7)), was how to determine what sort of signal to send so as best to convey messages of a given type. How could such messages be coded to secure the fastest error free transmission over a given circuit? The same problem exists in considering interpersonal communication; how can the difference between an intended and an interpreted message be minimised in conveying that message over the shortest possible time? In order to answer this question a number of factors have to be considered which can be elucidated in the information processing model (see Figure 1).

The intended message must first be encoded and physically transmitted in the form of a signal. In interpersonal communication this encoding and transmission is achieved by selective use of signs whether written, spoken or gestured. For example, the gestures of a policeman controlling traffic, the sentences, tables and diagrams used in written procedures, or a verbal instruction from a supervisor, all involve the use of signs which are transmitted in different forms. In effective communication the signal acts upon the receiver in such a way as to cause him to select the appropriate behaviour from a set of alternatives in the shortest possible time. There would be an accurate decoding of the signal if for example, the driver turns right in response to the policeman's right turn gesture; in other words, the policeman's selection of sign and the driver's selection of response match.

This encoding-decoding process involves some interesting problems. Natural languages such as English abound with ambiguity. While this allows for subtlety of expression, accurate communication requires a clearly defined context such that the encoded selection of signs by the sender unambiguously indicates the same selection by the receiver. The sender may transmit the message "Ring Janet", but the receiver may know a variety of Janets. The set of possibilities from which the sender selects "Janet" is therefore different to the set from which the receiver makes the selection, and because of this the latter is likely to select the wrong Janet. To reduce error, therefore, it is necessary to specify context or set by saying for example, "Ring Janet in the Ergonomics Unit" or by signing the message, or by any means which specifies the set to which Janet belongs. Although this may seem a rather trivial example it does illustrate the need to formalise communication procedures, where coding and decoding errors can be reduced if the process works within a well defined set of rules of selection resulting in an inflexible closed system (although rules may still be broken, of course). All organisations have rules but these rarely apply to communication systems. When communication systems are open and therefore flexible (unlike the closed traffic directing system used by policemen) they are more prone to error because the set of signs upon which they are based is not effectively finite; neither are the signs free from unwanted or changing associations. Take this example from Barrett (8) querying nuclear power plant communication procedures:

> "[The instructor] replied that there were no directions about communicating between operators - operators must develop their own communications system.... I heard one instructor ask another ... what the temperature was. "It is normal", came the reply. "Don't tell me whether it is normal," said the instructor ...,"tell me what it is."" (p.162)

Here, the intention of the question was unclear and therefore it was not possible for the operator to select accurately the type of answer wanted. This example also illustrates that in open communication systems, ambiguity in the signal could be reduced not only by limiting the sign set but also by enabling a two way communication process. In this way the sender can either correct the receiver's response or the receiver may ask for the context to be more accurately specified, such as, using the earlier example, by asking "Which Janet?"

Even if the sender and receiver do not make errors in the selective encoding and decoding of messages, all communication signals are prey to disturbances or noise. This noise is added to the message as bogus, unwanted information, and this affects the receiver's potential for making accurate selections and is likely to increase the time that this takes. For example, when relevant information is buried in a mass of irrelevant information, the latter constitutes noise. Such a situation significantly contributed to causing the serious accident at Hixon level crossing for example, (Great Britain: Ministry of Transport, (9)) which resulted in 11 deaths and 45 injuries. The Ministry of Transport did not emphasise important information in its national publicity concerning the operation of the new automatic crossings. Neither did British Rail, responsible for local publicity, who produced a weighty document containing a mass of irrelevant information as far as the disparate organisations who received it were concerned.

Fuchs et al (10) comment on similar deficiencies in written emergency procedures; either commands were written as added cautions and notes instead of specific statements, or extraneous explanatory information was included. This is an example of the latter:

> "Following a period of approximately constant temperature, the primary coolant temperatures increase to well above the secondary saturation temperature, indicating the loss of heat sink by dryout. This is an important indication of the approach to inadequate core cooling. If the RCP's are off, cold leg temperature will initially increase faster than hot leg temperature, resulting in decreased core /\T. Use TR-0115 and TR-0125 for cold leg temperatures and TR-0111 and TR-0121 for hot leg temperatures. Using the recorders will show a trend for easier determination of temperature trends." (Sec.2, p.13).

The authors (10) recommend the following remedy:

> "Using appropriate temperature recorders (see table), determine hot leg and cold leg temperature trends. Go to step indicated by table."

[A table of temperatures was provided] (Sec.2, p.14)

In the previous case, the explanation of the system was irrelevant; the superiority of the remedied procedure should be evident.

When communications are unavoidably subject to noise and therefore the receiver is likely to make an uncorrectable and unidentifiable mistake, the introduction of redundancy into the message can be used. All redundancy is, in effect, addition and in its simplest form is plain repetition. Messages exhibit redundancy if they contain less information than they could by virtue of an excess of rules; by implication, messages could be shortened by removing redundancy. This must be distinguished from the concept of irrelevant information mentioned earlier, since a message sequence which exhibits redundancy enables one part of the message to be predicted from knowledge of another, e.g. "Mary .... lamb, its fleece .... snow." The system of specifying code letters in air traffic control or in police radio communications is an example of use of redundancy, e.g. "Nectar one four metro" for NI4M. Similarly, the use of a whistle code accompanying the traffic control gestures of Italian policemen is also redundant.The term "redundancy" is therefore something of a misnomer whereas "irrelevant" clearly refers to information which is of no value in carrying out a particular task.

Besides noise, other factors affect transmission, for example the capacity of the communication channel(s). If channels are being used to capacity, additional signals will be lost or delayed until the channels are cleared. Busy telephone lines are a good example of a communication channel being blocked because of limited capacity. Alternatively, receiving several communications at once that exceed the processing capacity of the receiver can result in delay as response priority will need to be determined, and in some cases a complete failure to respond to certain messages will occur. Both these examples have occurred as contributory causes of accidents (e.g. Great Britain: Department of Transport, (11); Great Britain: Health and Safety Executive, (12)). Capacity problems may exist because resources are limited, or they may be exacerbated when the number of people in a communication network increases but the available channels of communication do not, or when the need for communication increases as can occur in emergency situations. Rules for communication also need to specify, therefore, not only <u>what</u> and <u>how</u> to communicate but also <u>to whom</u>. This point is elaborated in the following section.

## SPECIFIC COMMUNICATION PROBLEMS

Although the use of an information processing model enables error prone points in the communication process to be identified, an examination of actual communication errors can be used to highlight specific weaknesses in a particular system. Communication errors are often difficult to identify because the consequences are generally not immediately observable, if at all. For this reason much of the evidence must be taken from cases where accidents have actually occurred and although this may bias conclusions it is, at present, the only reliable available source of data.

### What to communicate

#### 1 Language set limited by lack of system knowledge

In the explosion accident at Houghton Main Colliery, Yorkshire (Great Britain: Health and Safety Executive, (12)) an electrician recorded a fault in a ventilation fan which had been switched off because it had been sparking. He

[A table of temperatures was provided] (Sec.2, p.14)

In the previous case, the explanation of the system was irrelevant; the superiority of the remedied procedure should be evident.

When communications are unavoidably subject to noise and therefore the receiver is likely to make an uncorrectable and unidentifiable mistake, the introduction of redundancy into the message can be used. All redundancy is, in effect, addition and in its simplest form is plain repetition. Messages exhibit redundancy if they contain less information than they could by virtue of an excess of rules; by implication, messages could be shortened by removing redundancy. This must be distinguished from the concept of irrelevant information mentioned earlier, since a message sequence which exhibits redundancy enables one part of the message to be predicted from knowledge of another, e.g. "Mary .... lamb, its fleece .... snow." The system of specifying code letters in air traffic control or in police radio communications is an example of use of redundancy, e.g. "Nectar one four metro" for NI4M. Similarly, the use of a whistle code accompanying the traffic control gestures of Italian policemen is also redundant.The term "redundancy" is therefore something of a misnomer whereas "irrelevant" clearly refers to information which is of no value in carrying out a particular task.

Besides noise, other factors affect transmission, for example the capacity of the communication channel(s). If channels are being used to capacity, additional signals will be lost or delayed until the channels are cleared. Busy telephone lines are a good example of a communication channel being blocked because of limited capacity. Alternatively, receiving several communications at once that exceed the processing capacity of the receiver can result in delay as response priority will need to be determined, and in some cases a complete failure to respond to certain messages will occur. Both these examples have occurred as contributory causes of accidents (e.g. Great Britain: Department of Transport, (11); Great Britain: Health and Safety Executive, (12)). Capacity problems may exist because resources are limited, or they may be exacerbated when the number of people in a communication network increases but the available channels of communication do not, or when the need for communication increases as can occur in emergency situations. Rules for communication also need to specify, therefore, not only what and how to communicate but also to whom. This point is elaborated in the following section.

## SPECIFIC COMMUNICATION PROBLEMS

Although the use of an information processing model enables error prone points in the communication process to be identified, an examination of actual communication errors can be used to highlight specific weaknesses in a particular system. Communication errors are often difficult to identify because the consequences are generally not immediately observable, if at all. For this reason much of the evidence must be taken from cases where accidents have actually occurred and although this may bias conclusions it is, at present, the only reliable available source of data.

### What to communicate

#### 1 Language set limited by lack of system knowledge

In the explosion accident at Houghton Main Colliery, Yorkshire (Great Britain: Health and Safety Executive, (12)) an electrician recorded a fault in a ventilation fan which had been switched off because it had been sparking. He very correctly reported the defect on his report sheet but did not note the fact that the fan was sparking. Had the fan later been repaired, this omission would have been of little consequence. However, it was not, and the firedamp that subsequently built up in the mine was later ignited by sparks from the unrepaired fan when switched on a few days later. The electrician was solely concerned with electrical faults, and the set of responses associated with this role would appear not to have included consideration of the implications of sparking with regards safety.

2 Language set limited by expertise

The limitations imposed by lack of knowledge or lack of consideration of the safe operation of a system in the above example, is also amply reflected in expertise. It is not surprising that the influx of hundreds of experts on the occasion of the Three Mile Island Accident (President's Commission on the Accident at Three Mile Island, (13)) generated an equally large number of interpretations of the nature of the accident (Nelkin, (14)).

Expert testimony is not always what a situation requires. Prior to the 1966 Aberfan disaster, where a coal tip descended onto the village killing 116 children and 28 adults, the National Coal Board had quelled the fears of the local council by using an expert. They sent the Area Planning Engineer to a Town Planning Committee Meeting. Unfamiliar with the situation and unbriefed, he advanced a tier-tipping scheme without giving it proper consideration, but nonetheless presenting it with an air of authority. The borough accepted the persistent reassurance from NCB officials who were, after all, legally responsible and presumably expertly qualified (Bignell, (15)).

3 Ambiguous and inappropriate messages

Communications can be ambiguous or inappropriate. In the Sellafield incident, Standing Instructions, January 1983, regarding highly active plant wash tanks, required that a solvent float off procedure be carried out. However, Plant Washout Instructions, October, 1983, were annotated with a handwritten note which said "For this shut down it is proposed not to float off and recover solvent from HAPW's." Although ambiguous and confusing, this note was not actually the cause of highly radioactive solvent being ultimately discharged into the sea. According to the reports (4) (5) the reason appears to be related to inappropriate information recorded in the shift log book on 10th November concerning the contents of highly active plant wash tank B (HAPW B). The records of contents were as follows:

"ejections from HASW [highly active solvent wash] (am shift 8th Nov.)
"ejections from HASW washes" (am shift 9th Nov.)
"ex HASW washout" (am shift 10th Nov.)

This last entry, later amended, suggested that solvent had been floated off. This was not the case, unknown to the afternoon shift, and discharge of HAPW B contents to Sea Tanks was authorised.

In a railway incident where a train was found to have defective brakes, a message to the Power Controller was interpreted as suggesting that the train should be changed, if possible. Lack of knowledge on the part of the receiver about the state of the train and a failure on the part of the sender to unambiguously state the serious nature of the brake fault led to the train remaining in service (11).

The person writing the procedures for the use of safety missile pins on

RAF jets failed to take into account that Phantom jets might be a special case. In fact they were impossible to fly with the pins in place. The result was that the procedures were ignored, a Phantom jet was used in a training mission with live missiles and a Jaguar aircraft was subsequently shot down at a loss of about £7 million (The Times, (16); The Guardian, (17)).

4 Sender-receiver mismatch

When communications occur across systems boundaries, problems in relating the sender's knowledge to that of the receiver's can result in error, such as the provision of irrelevant or "foreign" information and/or the omission of relevant information or emphasis, e.g. as occurred in the Hixon accident (9) mentioned earlier.

## How to Communicate

1 Means of communication not available

Although an operator may have information regarding a particular system state he may not actually be able to communicate this information because the means to do so are not available. An example is given by Hunns (18) concerning the use of a now outdated communication system between signalmen at either end of a tunnel. The system involved the use of two way communication between signalmen using needle telegraph, an automatic semaphore signal to train drivers, and a one way flag signalling system between signalman and driver in case of failure of the automatic system. Because the system was designed to prevent two trains being in the tunnel at once, it did not provide a means of communicating "two trains in tunnel." When this situation actually occurred, the subsequent misinterpretation of messages transferred between signalmen as a result of an inability to communicate the actual system state, resulted in a third train being allowed to enter the tunnel.

In the above example, communication about a particular system state was not possible. However, communication may be blocked because of limited resources. One example is of the chargehand fitter in the Queen Street station accident (11) who, because a telephone link was engaged, failed to contact the maintenance controller about the train's faulty brakes and subsequently allowed the train to remain in service. The maintenance controller would have put a one journey restriction on the locomotive had he received the call. As it was, he assumed the chargehand was still working on the locomotive and that no further action was required.

In the Phantom jet incident (16) (17) warnings that live missiles were being carried could not be provided. Both the red tape normally placed over the master switch when live missiles were being carried and the statutory yellow notices indicating "aircraft armed" were not available.

In the above examples personnel made assumptions about the system state. In all cases they were wrong.

2 Formal methods break down

The use of log books has frequently been a source of error. The Sellafield incident (4) (5) is a good example (see earlier description). This system not only relies on complete and accurate information being recorded but also on the necessity to do so being perceived by the sender and the necessity to actually read the message being perceived by the receiver. Logs seem to be used to transfer information between shifts or between groups with different functions. For example, train drivers may enter problems in a repair book for referral to maintenance departments, or operators may record notable incidents for subsequent inspection and analysis by management. Errors in this sort of record system are quite common, ranging from failures to make entries to loss of a record book. These errors result in reliance on informal communication systems such as passing on information by word of mouth or by notes (e.g. (12)), a system which can easily fail.

It may be that formalised means of communication come to be regarded as unnecessary by those who are familiar with using the system. In the Queen Street Railway accident (11) for example, fitters failed to acquire information from standing orders because they considered that learning from more senior fitters directly was a more efficient means of obtaining information regarding maintenance procedures. As a result standing orders relating to a new class of locomotives were not read, despite being made available in the library.

### Who to communicate with

#### 1 Interpersonal barriers

There is a great deal of evidence to show that communication barriers exist between people, not only because of differences in physical location, but also due to factors such as status or personality differences (e.g. Blau (19), Gaines, (20), Read (21)). Such problems exist in hierarchical organisational structures. Operators tend to ask their workmates for help rather than their supervisors for fear that this would reveal their incompetence. Downward communication flow in a hierarchy is less impeded and tends to be in terms of instructions or orders but is also reflected in the use of experts. However, upward communication is also important because those closest to the task are apt to have knowledge of facts and details unknown to superiors. The Aberfan disaster (15) is a good example of upward communication failure. At the time, NCB archives showed no record of similar previous tip slides. Also, warnings of sinkings by the chargehand at the Aberfan tip were largely ineffective. The downward process may also fail. Managers have been shown to be lacking in issuing sufficient safety warnings in some of the accidents mentioned.

#### 2 Information handling problems

Another problem evident in many of the accidents already mentioned is that complex organisational structures frequently involve the transference of information between a large number of people or between systems. The larger the number of people involved, the longer the message takes to reach its destination and the more prone to distortion or loss of information it is likely to be. Turner (22), found that inter-organisational grouping results in information handling difficulties, frequently a contributory cause of major disasters. In addition, informed outsiders can be considered to be uninformed alarmists, a clear example of organisational exclusivity.

## CONCLUSIONS

The use of an information processing model is not ideally representative of all the aspects of interpersonal communication. This is particularly so when a natural rather than artificial language forms the basis of the process. However, it does serve to draw attention to error sensitive areas in the process. Examples of errors drawn from real life situations demonstrate the need to consider a wide organisational context as well as the characteristics of senders and receivers who participate in the process, particularly their knowledge and perceptions of the system. Because this is such a poorly researched area, further conclusions are limited until more work can be carried out. In the meantime, it is hoped that those who have control over systems design and safety will make use of the evidence that is currently available. It is not only necessary to formalise the inter personal communication process, but also to ensure that such formalisations are not allowed to lapse. Also, by providing sufficient noise free transmission resources, where possible, particularly where abnormal system states are likely to occur, and by directing the flow of communication, communication structure weaknesses can be minimised.

Finally, it should be apparent that natural languages are a highly imperfect means of communication as far as system safety is concerned. Future research should perhaps concentrate more on designing artificial language systems which are not prone to the problems of the alternative open systems currently in use.

## REFERENCES


1. Wright, P., and Barnard, P., 1975, Applied Ergonomics, 6, 213-220.

2. Losee, J.E., Allen, R.H., Stroud, J.W., and Ver Hulst, J., 1962, "A Study of the Air Force Maintenance Technical Data System", Aerospace Medical Research Labs., WPAFB, Ohio, Report No. AMRL-TRD-62-85.

3. Brune, R.L., and Weinstein, M., 1981, "Checklist for Evaluating Emergency Procedures used Nuclear Power Plants", Nuclear Regulatory Commission, U.S., Report No. NUREG/CR-2005.

4. Great Britain: Department of the Environment, 1984, "An Incident Leading to Contamination of the Beaches near to the British Nuclear Fuels Limited, Windscale and Calder Works, Sellafield, November, 1983, HMSO, London.

5. Great Britain: Health and Safety Executive, 1984, "Contamination of the Beach Incident at British Nuclear Fuels Limited, November, 1983", HMSO, London.

6. Bellamy, L.J., 1983, "Neglected Individual, Social and Organisational Factors in Human Reliability Assessment", Fourth National Reliability Conference Proceedings "Reliability '83", Birmingham, England.

7. Shannon, C.E., and Weaver, W., 1949, "The Mathematical Theory of Communication", University of Illinois Press, Urbana, Illinois.

8. Barrett, R.S., 1982, pp. 161-171 in C.P.Wolf and V.B.Shelanski (Eds.) "Accident at Three Mile island: The Human Dimensions", Westview Press, Colorado.

9. Great Britain: Ministry of Transport, 1968, "Report of the Public Inquiry into the Accident at Hixon Level Crossing on January 6, 1968", HMSO, London.

10. Fuchs, F., Engelschall, J., and Imlay, G., 1981, "Human Engineering Guidelines for use in preparing Emergency Operating Procedures for Nuclear Power Plants," Nuclear Regulatory Commission, U.S., Report No. NUREG/CR-1999.

11. Great Britain: Department of Transport, 1980, "Railway Accident: Report on the Collision that Occurred on 24th December 1977 at Queen Street Station in the Scottish Region British Railways", HMSO, London.

12. Great Britain: Health & Safety Executive, 1976, "Report on the Explosion at Houghton Main Colliery, Yorkshire", HMSO, London.

13. President's Commission on the Accident at Three Mile Island, 1979, "The Need for Change: The Legacy of TMI", Pergamon Press, New York.

14. Nelkin, D., 1982, pp. 143-151 in C.P.Wolf and V.B.Shelanski (Eds.) "Accident at Three Mile Island: The Human Dimensions", Westview Press, Colorado.

15. Bignell, V., 1977, pp. 37-65 in V. Bignell, G. Peters and C. Pym (Eds.) "Catastrophic Failures", Open University Press.

16. The Times, 1983, 10th Jan. p.1: 11th Jan. p.1; 12th Jan. p.6; 13th Jan. p.6.

17. The Guardian, 1983, 11th Jan. p.1 and p.3; 12th Jan. p.2

18. Hunns, D.M., 1982, "Psychology of Communications", Proceedings of "7th Advances in Reliability Technology Symposium".

19. Blau, P.M., 1955, "The Dynamics of Bureaucracy", University of Chicago Press, U.S.

20. Gaines, J.H., 1980, <u>Human Relations, 33</u>, 929-942.

21. Read, W.H., 1962, <u>Human Relations, 15</u>, 3-15.

22. Turner, B.A., 1978, "Man-Made Disasters", Wykeham Publications, London.

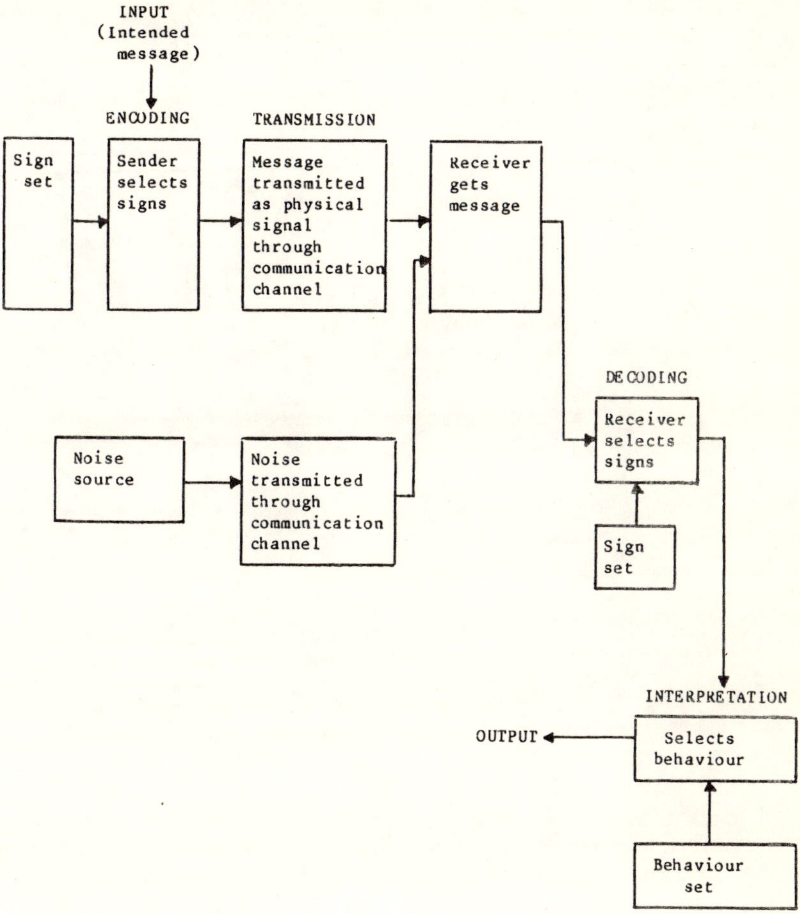


Figure 1 An information processing model of human communication.

# AUTOMATION AND THE HUMAN OPERATOR; EFFECTS IN PROCESS OPERATIONS

Alfred A.F. Brouwers*

In the past decades automation has had a significant influence on the nature of operator tasks in the process industries. General lines of changes in regard to task, organization and health and well-being of operators are reviewed. In particular with regard to tasks of operators the lines of changes are extended into the near future. In the prevention of bottlenecks in the tasks great importance has to be attached to the design process of an automated man machine system.

## 1. INTRODUCTION

In the past decades important changes have come about with regard to process operations in the process industries. On the one hand advanced control equipment has made possible the automatic control of complex processes. On the other hand further advances in control equipment were needed as a result of a tendency to large scale process systems, increased efficiency and safety and the increasing importance of environmental factors.

As a consenquence of these changes the role of the human operator has changed. First the operator had direct visual contact with the 'pots and pans'. By means of e.g. handwheels the system was also controlled on the spot. As pneumatic, electric and later electronic devices became available monitoring and control actions were carried out from a remote, central control room. By this not only the physical working environment has changed. In particular the tasks of the operator have been influenced. In short there has been a shift from (local) manual control of a small (sub)system to (remote) supervisory control of a large integrated system.

Several studies (references 1, 2) have indicated that this shift may give cause for problems to the operators as a consequence of bottlenecks in the work and the work conditions. Such problems may find expression in too low or excessive task-load, demotivation, sickness absenteeism, turn over, etc. This may act in an unfavourable way upon the task execution and therefore upon the course of events of the process.

The above-mentioned is reason to review a number of significant changes and related effects with regard to the operator and his task situation.

* Netherlands Institute of Preventive Healthcare-TNO, Leiden, the Netherlands

The above is also reason to extend the general lines of these changes over the past decades into the near future. Of particular importance is the question as to what bottlenecks can be expeçted in the near future task situations of the operator. The prevention of bottlenecks in the tasks allocated to man is starting-point for a reflection on the role of the design process of an automated man machine system (MMS).

## 2. TECHNOLOGY, 1940-1980

In automated process operation mechanization can be considered as a prerequisite. The second condition is that the system has a receptive faculty to information with regard to process variables that have to be controlled. It follows as a matter of course that this information can be transmitted, processed and displayed.

In the period after the second world war the possibilities to process information (key-stone to automated process operation) have increased enormously. To a large extent this rapid growth has become possible by the advancement in electronics starting in the fifties. The availability of ever smaller components with still increasing capacities has made possible the construction of complex information processing systems. Improvements in the ratio between capacities and prime costs have played a prominent part in this ($10^5$ in ten years; reference 3). Scheme 1 presents an overview of developments in control equipment, signal transmission and the structure of the man machine interface.

SCHEME 1 - Technological developments; 1940-1980

| | | '40-'50 | '50-'60 | '60-'70 | '70-'80 |
|---|---|---|---|---|---|
| electronics | components | transistors | Integrated Circuits (IC's) | micro-processors | single-chip computers, submicron-technologies |
| | integrated systems | computers with tubes | computers with transistors | computers with IC's | |
| control equipment | | handwheels, valves | remote-control, first applications proces-computers | digital instrumentation, redundancy, proces-computers | local control by digital instrumentation |
| signaltransmission | | pneumatic, mechanical | pneumatic, electrical | electronic, pneumatic | electronic, pneumatic |
| man-machine interface | | local panels with large analogue displays | complete centralization in control room, analogue displays | centralization, smaller interfaces, digital instrumentation, first applications of visual display units | decentralization, miniaturization of interfaces |

## 3. AUTOMATION AND TASK OF THE OPERATOR

In the process industries control and monitoring are important tasks of the operator. In the past decades the progress of technology has made it more and more possible to allocate elements of the control task to the machine component of MMS. Although in a more indirect way it has also become possible to allocate aspects of the monitoring task to the machine.

The advance of technology has been described in literature in several scales for levels of automation (references 4, 5, 6). Although very useful these scales are too detailed to be used in this overview. Here, the degree of automation is conceived as a concept that indicates to what extent the control or the monitoring task has been allocated to the operator or to an automatic device. A high degree signifies that the activity is executed to a large extent by the automatic device. Essentially less of the activity remains then to the operator.
In coherence with scheme 1 three levels of automation of the control task are distinguised (the levels refer to a stationary course of events in the process).

Level I: Manual control

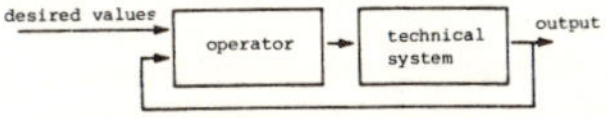


existing configuration '40-'50

The operator forms part of the closed loop of the control of the system. He performs control actions in order to minimize deviations between output and desired values.

Level II: Automatic control

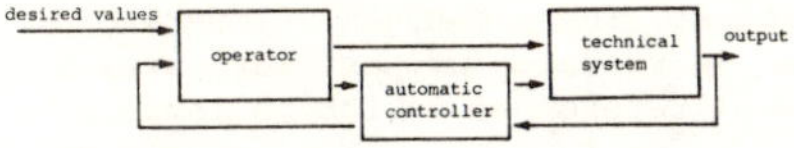


initial applications '50-'60

In this configuration the operator as well as the automatic controller execute control actions in order to minimize deviations between output and desired values.

Level III: Supervisory control

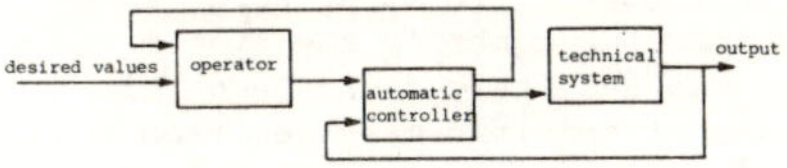


initial applications '60-'70

In this configuration the system is controlled by the automatic controller. The operator forms no more part of the closed loop. His task is mainly a monitoring one. The operator can interfere only by setpoint-control.

Based on the frequency and the purpose of the observation of display information a distinction is made between 3 levels of automation of the monitoring task (also with regard to a stationary course of events in the process):

- Level I: Intensive monitoring. The operator observes the displays almost permanently. The information is needed for the control of the systems.
- Level II: Active monitoring. Monitoring activities aim a a regular inspection of the conditions and operations of the process. The operator takes the initiative; control actions may result from the display-inspection.
- Level III: Survey monitoring. Surveying the process situation and - trends in - the operations is now characteristic to the monitoring activities. Often the activity is prescribed and coupled to administrative work. Control activities very seldom follow on the monitoring actions.

The levels of automation of control and monitoring can be used to set up a classification of operator-tasks (see figure 1).

FIGURE 1 - Classification of operator tasks with regard to the level of automation of control and monitoring

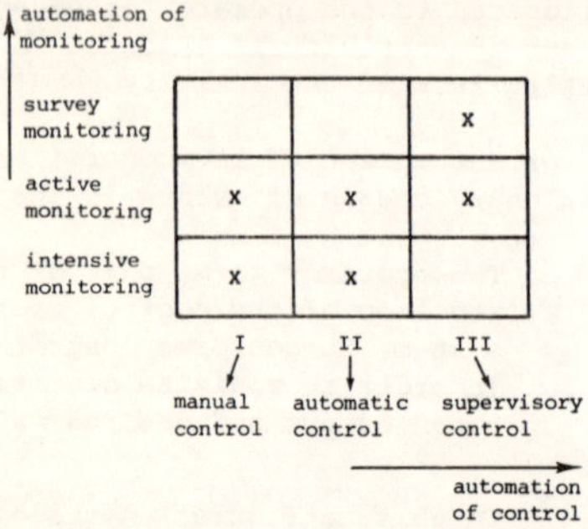


In figure 1 a number of cells contain crosses. These concern 19 operator tasks in particular in the process industries. (A research project was carried out here by the NIPG; see reference 2; see also 'reference' 14 for an indication of the 19 task situations.)

The 19 systems are constructed as well in a recent as in an earlier point of time. As such the cells with crosses contain an aspect of technological development. On the task level the effect of this can be seen in a diagonal bandwidth from the lower-left to the upper-right in figure 1. In other words technology has changed the situation in which the operator controls the system and in doing so permanently needs information. Now the automatic device controls the system. The operator interferes only in case of specific messages from the system. The monitoring task has become more and more of a passive nature.

These developments have not been without consequences for the subjective experience and health and well-being of the operators. Based on the afore mentioned research an overview of these aspects is presented in the next paragraph.

## 4. AUTOMATION, WORK EXPERIENCE AND HEALTH

Research in the fifties and early sixties gives a quite favourable picture of the work in automated process systems for example in comparison to work in mass-production systems, e.g. assembly-lines (references 7, 8, 9). The work is more interesting, challenging, offers more variation, etc.

In later research, early seventies, there is doubt that each higher level of automation will still result in such an image of the work and work situation (reference 10). To investigate this a survey was carried out in 19 different task situation with different levels of automation of the control task of the system. Operators (N=186) have been interviewed with regard to various aspects of the work, health and well-being.

Tabel 1 presents an overview of a number of significant results of this project. (For reasons of surveyability data are presented in a qualitative way.) The degree of automation in tabel 1 corresponds to the levels of automation of the control task in paragraph 3. In the project it appeared that with regard to level II a distinction had to be made in situations in which control is mainly by the operator (level IIa, e.g. air traffic control) and in situations in which the major part of the control actions are performed by the automatic controller (level IIb, e.g. a nitric acid plant).

TABLE 1 - Overview of research data of 19 different operator task at different levels of automation of the control task (++ = high score, + = relative high, = = average score, - = relative low, -- = low)

| | | | | | |
|---|---|---|---|---|---|
| system | level of automation | I | IIa | IIb | III |
| | system complexity | -- | -- | + | ++ |
| | system size | -- | - | + | ++ |
| task | % time controlling | 21.9 | 6.2 | 3.8 | 2.6 |
| | % time monitoring | 69.7 | 43.3 | 48.6 | 44.6 |
| | % time administrative work, etc. | 0.3 | 41.2 | 36.3 | 31.8 |
| | % rest | 8.0 | 9.1 | 11.5 | 21.1 |
| | routine control activities | + | = | - | - |
| | amount of information | + | + | - | - |
| | complexity of information | - | - | + | + |
| | feeling of control ability | ++ | - | - | = |
| | uncertainty in control actions | -- | + | - | - |
| | workload due to system disturbances | = | = | = | - |
| organization | size of team | - | - | + | ++ |
| | necessity of teamwork | - | - | + | ++ |
| subjective experience of operators | job satisfaction | + | + | - | -- |
| | achievement | + | + | - | - |
| | variation | + | + | - | - |
| | subjective workload | + | + | - | - |
| | satisfaction with teamwork | = | - | + | + |
| | autonomy | ++ | + | + | - |
| | attitude towards shiftwork | ++ | = | + | -- |
| | subjective well-being | + | = | + | - |
| | experienced health | - | + | + | - |
| | sickness absenteeism | - | + | + | ++ |

As the data of table 1 indicate the allocation of tasks between man and machine is less favourable at the higher levels of automation from the point of view of human capabilities and needs. The situation arises that there is a lack of process related activities for the operator, he is insufficiently involved in the operations of the process. This may evoke the feeling that he can no longer adequately control the system, he may feel uncertain in case of manual control actions. The result can be a situation in which the operator no longer feels that he can contribute in a meaningful way to the process operations. As a monitor he waits for irregularities in the process.

At the higher level of automation of the control task there is also a more negative appreciation of job satisfaction by the operators. The same applies to the results with regard to experienced health and subjective well-being. Sickness absenteeism (frequency) is higher at this higher level of automation.

The work in a usually larger team of operators is reason to contentment at the higher automation level. At the same time however, there is a loss of autonomy. At the lower level of automation the operator is more independent. On the opposite side there is a tendency that hierarchical higher functionaries intervene in the control of the process in case of disturbances, start-up and shut-down. Finally, at this level the inherent disadvantages of shift-work are more pronounced. This is due to a higher incidence of continuous shift work.

In general, from the data of table 1 there is reason to suppose that the positive appreciation of the job satisfaction in the process industries in the fifties and early sixties has not continued into the seventies. There even is reason to suppose that there is a tendency to the opposite. Is is a fact that for the time being the systems can be characterized as man machine systems. A complete replacement of man, so-called unmanned plants, seems not likely to take place in the near future. The responsibility of man for an efficient and safe process operation will also be operative in the near future.

## 5. FUTURE TASK SITUATIONS

### 5.1 Aspects of Technology

In the future the process industries will make use of the continuation of the advancement of control and monitoring equipment. The necessity to master costs of energy and raw materials, the importance of safe process operations and the enlargement of environmental laws are strong incentives for this. In the systems the conventional equipment will more and more be replaced by microprocessors and on-line process computers (we will use the expression 'process computer' or 'PC'). With regard to the use of PC among others the following possibilities can be distinguished (reference 11):

- planning and programming at the level of the enterprise;
- refining the control of process variables within critical limits;
- start-up, shut-down, alarm and safety procedures;
- quality control, quality measurements;
- programmed control of batch operations;
- (increased) use of visual display units.

Another striking development is the increased effort to enhance the reliability and availability of the PC, i.e. the construction of redundant systems with regard to hardware and software and technics for fault detection and correction (fault tolerant computers). Only recent such systems, by low costs and volume of components, have become economically feasible.

### 5.2 Aspects of Tasks and Organization

The question is what does the increasing PC implementation signify to the tasks of the operator. It is obvious that the afore mentioned possibilities are concerned as well with the control tasks as with the monitoring task. Therefore higher levels of automation of these tasks can be expected in the near future.

With regard to the control task it can be expected that next to situations in which there is a stationary course of events in the process, also special situations can be controlled at the level of supervisory control. For example: start-up procedures, the correction of a number of - well known - disturbances, control actions to bring a system back to safe-operations. Such a level of automation of the control task is indicated here as: extended supervisory control. Next to that it is conceivable, although in the process industries not at short notice, that *all* process situations in a system that are thinkable can be controlled at the supervisory level. As an example, outside the process systems sphere, we can refer to missions of the U.S. Space Shuttle in which five computers control the entire flight (reference 12). For such a level the description 'extra supervisory control' is proposed.

With regard to further automation of the monitoring task the increase of reliability and availability of the equipment seems of particular importance. This may bring out situations in which the operator does not execute the (prescribed) 'survey monitoring', but is peaged by the system whenever that is necessary. At such moments the systems provide relevant information and, the operator can request addional information. For such a level of automation of the monitoring task the indication 'monitoring by exception' is proposed.

Figure 2 presents a classification of present and near future operator tasks with regard to the level of automation of control and monitoring.

FIGURE 2 - Classification of existing (shaded) and future operator tasks with regard to level of automation and monitoring

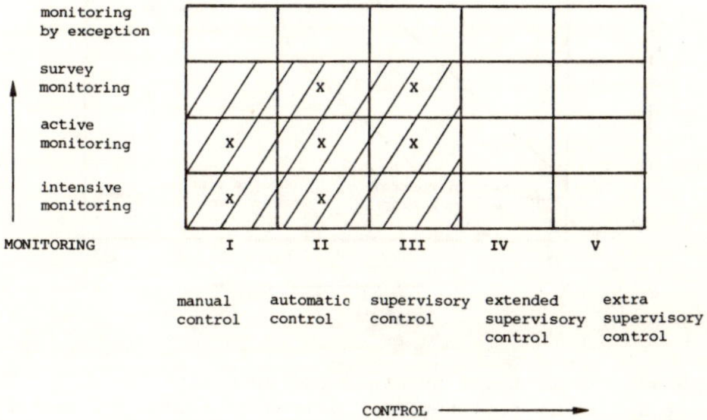


On the one hand the question can be raised whether there are or will be appropriate task situations for all cells in figure 2 (e.g. intensive monitoring x supervisory control). On the other hand it can be said that future technology provides eleborate conditions to adjust the automation of control and monitoring to the characteristics, the demands of a specific system. As such operator task situations may become considerable different in nature. In case the automation of control is enlarged, then the operator is going to supervise the computer. In case the automation of the monitoring task is emphasized, then the computer supervises the operator. Both combined brings out the situation of 'computer control x computer monitoring'. The several possibilities are illustrated in figure 3.

FIGURE 3 - Dimensions of automation- control and monitoring. Possible different levels of automation of the control and monitoring task
* present trend

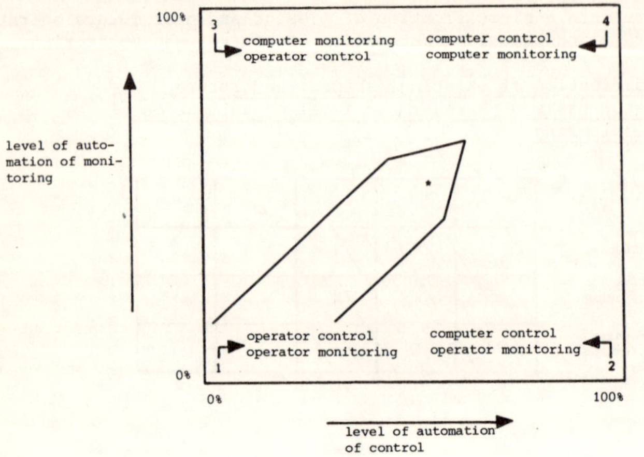


One of the aspects that can be mentioned here with regard to future organizational situations is that new specialized jobs will be needed in the control room. For example, the maintenance and management of hardware components and/or software packages. Other specific tasks may be allocated to the operator, e.g. the automatic analysis of product samples. These two examples also indicate the difficulty in regard to a statement on the number of people needed in the future control room (although an increase can hardly be expected).

With regard to the formal structure of the team of operators no significant changes are to be expected. As over the past decades there will be a limited number of hierarchical levels (2 to 3) within the team of operators. Another question is whether there can be a rotation of operators over the different jobs in the team. Specialized jobs do not increase the possibilities for job rotation. Finally, it can be expected that accomodations of shift schedules can decrease the inherent disadvantages of continuous shift-work.

### 5.3 Operator Tasks, Bottlenecks in the near Future

The previous discussions have concerned some developments of automation of control and monitoring over the past decades, and possible developments in this respect in the near future. The question must be raised what problems or bottlenecks with regard to these tasks can be expected in the near future?

We will indicate some of the possible bottlenecks in a number of questions with regard to control and monitoring in near future systems. The questions are focussed on man's role in situations of extended supervisory control x survey monitoring. In a way such process situations do exist already. An increasing number of task situations will 'reach' these levels in the near future. As both tasks are interrelated, specific questions may belong to the control as well as the monitoring questions.

*Questions with regard to control; extended supervisory control*

1. Will the ability of the operator to control the system remain at an adequate level? In this respect, it is necessary to control subsystems from time to time manually? Is simulator training an alternative in this respect? General issue: control skill, training of control skill.
2. Is it possible for the operator to make a prompt change over from survey monitoring to a more active way of system control? General issue: lack of balance in the workload and time pressure.
3. What must be the form of the interaction between man and the (extended) automatic controller, e.g. protocols of communication for man-computer dialogue? What alterations in system operations have to be passed on to the operator? When has the system to indicate motives for alterations? General issue: man machine communication.

*Questions with regard to monitoring; survey monitoring*

4. Is it possible for the operator to maintain an adequate level of alertness? Does performance on the monitoring task degrade as a function of time? General issue: possible effects of vigilance in monitoring complex system.
5. More than in the past process data can be expressed in digits. The possibility to compile information has also increased. Usually such data are not presented simultaneously but in successive requests by the operator. The question is what has to be the structure of the search process of (integrated) information. Has there to be a hierarchy, how many levels? General issue: man-machine communication.
6. What is the usefulness of e.g. a simplified process flow diagram of a complex and large system? What is the usefulness of a number, a hierarchy of such diagrams that can be presented on a visual display unit? Have such diagrams to be composed in a more abstract way or in a topographic way? General issue: the structure of overview information (in particular with visual display presented data).
7. What are the optimal characteristics of future process alarm systems? What are the conditions of information presentation for an adequate fault management by the operator? Does the presentation mode affect the detection, diagnosis etc. of system disturbances? For example, sequential visual display presentation may cause problems in periods of high signal density (reference 13). General issue: characteristics of the process alarm system and fault management.

It can be deducted from the afore mentioned that the near future allocation of tasks between man and machine may become even less favourable from the point of view of human needs and capabilities (e.g. an increasing lack of balance in the workload). The number of situations in which interventions in the process operations by the operator is needed will further decrease. To the operator the complexity of such situations however may increase (control skill, man-machine communication). More and more the situation will be that the operator - as a monitor - has to wait for irregularities in the system operations (vigilance, overview information). It is also obvious that there is a good deal of questions with regard to an adequate presentation of process data for such rare occasions (process alarm system, fault management).

Another aspect is that at various locations in the process industries the requirements with regard to professional training do increase. This may add to the discrepancy between what people expect of their job and what every day practice can offer.

In general it can be said that the more negative appreciation that started in the late sixties - early seventies may very well continu into the near future. It cannot be said that there is a predominant trend towards an operational symbiosis between man and machine. Man may even be put in the position of a weak link in a chain. Such a position, however, is opposite to the resonsibilities of the operator in a man-machine system.

## 6. THE DESIGN OF A MAN-MACHINE SYSTEM

A general accepted principle in methods of system design is a so-called phasic structure: system analysis, - design, - implementation and system operation. The activities in a preceding phase result in data that are used in the following phase. For example, system analysis implies the organization of the design process, the definition, goals and requirements of the system. The second phase brings out the actual design. This is realised in the third and put in active operation in the fourth phase.

It is evident that in each phase a number of choices and decisions has to be made. In regard to the nature of such decisions and choices the following can be stated.

The process of the design of a technical system often will make use of models of system operations. As the systems often are of a physical nature, mathematical models can be set up. Such models enable a detailed analysis and evaluation. Decisions then can be based on a quantitative way of weighing system design concepts against system requirements. Another aspect is that the design of a technical system customary is an engineering activity. Solutions to problems and bottlenecks in the system that is designed, are mostly found in technical rearrangements of the system. For example, an acceptable level of control of process operations in large scale systems has become possible by improvements of the control equipment.

As such in the design of a technical system a quantitative, technical approach follows more or less automatically. However, as long as man in MMS has to perform specific functions, and even has to comply with important operational events, the role of man also has to be taken into account in the design process. Problems and bottlenecks in the process control of an automated MMS are not merely of a technical nature. Criteria used in the design process therefore also have to aim at the human function in MMS.

In literature schemes of the design process have been presented that indicate that the machine component of MMS has to designed in parallel with the 'man component'. Such an approach is proposed as a better guarantee that the technical system, the task and the organization are designed as a coherent entity. The improved balance between man and machine is believed to bring out an improved function of MMS. At the same time this integrated approach may induce a reduction in system development time and - costs. The process should be taken up by a multidisciplinary team. Preferably operators should be represented as well, in particular when they have experience with comparable systems.

A first question is whether the integrated course is pursued frequently in actual system design. Regarding this no data are available. From the number of publications over the past decade, in which the integrated design is proposed, it cannot be supposed that such an approach has a widespread use in practise. (Personal experience in industry affirms this assumption.)

A second aspect is that not in all cases a joint optimization of the technical, economical and social system is feasible. From the point of view of environmental law technical modifications of a system can be required. To the operator such modifications may aid to bottlenecks with regard to information presentation, fault management etc.

A third aspect is that data with regard to the man component are of a different nature. Different from the technical sciences human behaviour to a large extent cannot be expressed in quantitative criteria, in mathematical models.

Notwithstanding these aspects, questions with regard to the human function in the control room can be answered from expertise of the sociale sciences, from general ergonomic pinciples. With regard to e.g. task variation, work rest schedules, workload etc. critical values can be indicated. An analysis of a proposed task situation may lead to statements as 'workload too high, too low, acceptable'. In specific cases it is also possible to deduce a number of characteristics of the human task situation from the technical design in order to 'test' whether the demands of the human function are met. Other questions can be answered by data collected on similar systems. Finally there is the possibility of collecting new data, for example based on simulator studies.

The afore mentioned does not exclude that data, statements with regard to the human function will remain of a more or less qualitative nature. The integrated approach, however, one has to contrast with a situation in which no systematic attention is paid to the role of man in automated man-machine system.

Automation is like a number of different, but all fast moving vehicles. We are on those vehicles and we cannot jump off. Then there should be no hesitation to use avaible knowledge in order to provide man with the tools to control a vehicle's movements.

## ACKNOWLEDGEMENT

The author would like to thank his colleagues Dr. C.L. Ekkers and Ir. C.K. Pasmooij and Ir. P. Bockholts, Ir. A.A.M. Kop, Dr. Ir. W. Veldhuyzen and Ir. J. Vermeulen in coöperation with whom research projects were carried out on which this article could be based.

## REFERENCES

1. Lees, F.P., 1974 "Research on the Process Operator", in Edwards & Lees, 1974 "The human Operator in Process Control", Francis & Taylor, London, England.

2. Ekkers, C.L., Brouwers, A.A.F., Pasmooij, C.K., and Vlaming, P.M. de, 1980 "Menselijke Stuur- en Regeltaken", NIPG-TNO, Leiden, the Netherlands.

3. Abelson, P.H., and Hammond, A.L., 1977 "Electronics: the continuing Revolution", publ. 77-4, Amers.Ass.Advanc.Sci., Washington, USA.

4. Amber, G.H., and Amber, P.S., 1958 "The Anatomy of Automation", Electric. Manufacturing 78-83, 124-34.

5. Bright, J.R., 1958 "Does Automation raise Still Requirements?", Harvard Buss.Rev. 4, 85-98.

6. Kern, H., and Schumann, M., 1970 "Industrie-arbeit und Arbeiterbewusstsein", Europäische Verlagsanstalt (Studienreihe Soz.Forschungsinst. Göttingen (SOPI), Frankfurt, W. Germany.

7. Mann, F.C., and Hoffman, L.R., 1956 Individual and organizational Correlates of Automation", J.Soc.Iss. 12, 7-17

8. Blauner, R., 1964, "Alienation and Freedom; the Factory Worker and his Industry", Univ.Chicago Press, Chicago, U.S.A.

9. Brenninkmeyer, G., 1964 "Werken in geautomatiseerde Fabrieken", Swets & Zeitlinger, Amsterdam, the Netherlands.

10. Susman, G.I., 1972 "Process Design, Automation and Worker Alienation", Industr.Relat.11, 34-45.

11. Rijnsdorp, J.E., and Roffel, B., 1980 "On-line Computers in de Procesindustrie (1) gebruik van on-line Computers", Polytechnisch tijdschrift/ Procesindustrie 35 nr. 4, 189-94.

12. Titulaer, G., 1982 "De Computer als Procesbewaker. Uitval is duurder dan redundantie", Intermediar 18-26, juli, 2.

13. Kragt, H., 1983 "Operatortasks and annunciator Systems. Studies in the Process Industry", Ph.D. Thesis, Eindhoven Univ.of Technology, Eindhoven, the Netherlands.

14. The 19 task situations were located at:

Level I - hopper suction dredge, coating systems car factory;

Level IIa - centres for management of vessel-traffic (2x), air-traffic, underground, tram/bus-traffic, power distribution;

Level IIb - nitric acid plant, penicillin plant, cheese-factory, cement furnace, electricity generation (exper.plant);

Level III - ethylene plant, butadiene plant, centre for distribution of natural gas, electricity generation (2x), blast furnace.

# SOME REMARKS ON THE PROCESS OPERATOR AND HIS JOB
# THE HUMAN OPERATOR IN THE CONTROL ROOM OF TOMORROW

H. Kragt and M.J.M. Daniëls*

In this paper the operator-process situation is considered from a non-ergonomic point of view. Recruitment, selection, education and training of process operators are dealt with after discussing the position and role of the operator in the organization. Subsequently operator job requirements are discussed. Process simulation as a work sample in operator selection is advocated.
The design of the operator's job is discussed by applying criteria for a meaningful job. Job design in the control room of tomorrow is dealt with.

## INTRODUCTION

The human operator in the control room of a process industry exemplifies a man-machine system. Like all man-made systems, control rooms have never been ideally designed. For the optimum design of this man-machine system, two approaches can be distinguished:

1. One starts with the operator and then tries to fit the "machine" (interface, work situation) to the man. This is known as the ergonomic approach.
2. One starts with the machine and then tries to find the right man by recruiting & selection or one may need to fit the man to the machine by education & training.

The first author has evaluated several control rooms from an ergonomic point of view. From those evaluations advice on new control-room designs was given (Kragt & Piso (1)). During the research project which started in 1976, we grew more interested in the operator as a person and in the other approach (recruitment & selection; education & training).

It is striking, that in the psychological literature so little attention is given to recruitment, selection, training and performance appraisal of process operators. One reason could be the lack of "feeling" for jobs in such sophisticated technical settings. The lack of a well defined performance criterion could be another reason that few researchers dare to explore this area.
In order to supplement the thoughts on the second approach, twenty people chosen from personnel and training staff members, and plant managers of three big companies were interviewed alone or in couples in the spring of 1983. (Two oil refineries, two petrochemical industries and four chemical industries were involved.)

*Dept. of Indus. Engin., U. of Technology Eindhoven, The Netherlands

A questionnaire was used that they had received in advance. In total 11 interviews were sent for verification to the interviewees. In the following, the interview data as such are not reported. The authors are presenting their personal opinion; where necessary, the opinions of the interviewees are given.

## THE OPERATOR'S JOB: DESCRIPTION AND REQUIREMENTS

### Job analysis: scarcity of studies

Ideally job requirements should originate from task analysis. The authors do not know of any operator studies in which such an analysis was carried out. In the following paragraph an attempt is made to formulate in general terms the operator tasks. The classification of Rijnsdorp & Rouse (2) is used:
A. General inspection. The large amount of information must be analysed critically. With regard to a possible action, Boot (3) remarks that if such an action is not carried out quickly and professionally, the value of the process computer will go down rapidly.
B. Process control. The operator has to keep on carrying out actions such as: switching on and off, stabilising control (for loops on "manual"), trimming of setpoints (e.g. for product quality control), optimising control.
C. Off-normal handling. Such situations require a quick reaction. The operator has to detect, identify, interpret, compensate if possible and to correct.
D. Reporting & consultation.
E. System evaluation and improvement.

### Philosophies and policies

Opinions with regard to the precise task of the operator can be placed on a continuum with the following extremes:
The one extreme. The operator has to control one plant skilfully. He must obey the appointed rules strictly. When danger occurs the plant is shut-down and other people come to restart it. In principal anyone must be able to learn the "job". All optimum control of the process is in the hand of management. The operator needs no chemical, physical and technical background-knowledge. In planning a new plant, one would prefer to recruit new personnel that can be programmed for the new plant.
The other extreme. The operator is fully responsible for starting-up, controlling and shutting-down several plants. He contributes to discussions with staff and maintenance people as a fully responsible member of the team. By doing so, he increases his knowledge and understanding of the processes. He is able to interpret the information and can contribute to a better control of the process. He can be placed in other control rooms and can easily be trained in new information technology. Cognitive and social skills are required. The organization itself is characterized by the presence of only few hierarchical levels.

### Job requirements

Workers and management were asked, which features of behaviour and attitude they think are necessary for a good operator performance. This direct approach to collecting this information is mentioned by Tenopyr et al. (4). In this section we focus on trainees from outside the company who are recruited as operators.
A. Intellectual level and school education. In the Netherlands, new trainees used to be recruited from the pupils with intermediate general continued education and those from elementary technical schools. It appears, that these people are equally good for operator training.

B. Specific abilities and skills. The following distinction is proposed by the authors:

- "Technical feeling". This is a requirement, because the operator has to work with compressors, pumps, centrifuges and so on.
- Analytical ability. The integrated process units and the VDU-based information presentation require flexibility in thinking and understanding; the ability to analyse problems and to supervise complex situations.
- Skill in reporting. An important task is the exchange of information, both orally (telephone, walkie-talkie) and in writing (log-book, the most important means of communication between the crews).
- Skill in arithmetic. Sometimes calculations have to be made in control rooms; also computer-output must be verified.

C. Physical demands.

- Sensory-motor. The importance of the manual skills steadily decreases. Manipulating knobs, thumb wheels and key boards is not difficult. Of course, normal demands are made on the ears (audible alarms) and eyes (e.g. no colour blindness).
- Health. Work outside the control room (e.g. climbing up and down outside the building in extreme temperatures) makes its own demands; just as shift-work does.

D. Personality traits. In literature personality traits are mentioned that would differentiate between a good and a less good operator (Hiscock (5); Brenninkmeijer (6); Lees (7)). Our interviewees also agree on the importance of personality traits. The ideal operator is described as conscientious and solid ("with sense of responsibility"); reliable and loyal; self-disciplined and balanced ("proof against panic"); characterized by self-confidence and authority; a sociable person. The question arises what functions would not need such requirements.

## RECRUITMENT & SELECTION

### Recruitment

In the Netherlands the aspirant operators are generally recruited from: 1. pupils with intermediate general continued education and those from elementary technical schools; 2. candidates with a particular professional experience.

ad 1. The interviewees suggested, that pupils with only primary school background should no longer be recruited. The complexity of most of the plants requires an amount of intelligence, studiousness and flexibility, that cannot be expected in this category of people. Moreover, it would be necessary to instruct them in arithmetic, the Dutch language, mathematics and physics to bring these subjects up to a suitable level, which would take time and cost money.
Some interviewees suggested, that economic recession has placed the companies in a favourable position. They can choose the best on the labour market. However, we have to warn against over-selection. The average operator job does not require a graduate.

ad 2. Among the candidates with a professional experience, it appears that ship-mechanics best meet the requirements. Their former job resembles that of the operator. They have a broad technical knowledge and experience; are used to "shift"-work and team work in difficult situations. The ex-mechanics have no difficulty in carrying out their job and soon rise to executive level.

We have to keep in mind, however, that education, professional experience and age are mutually contaminating factors; sometimes one regards professional experience as the main factor, while age or educational level are also concerned.

## Selection

Recruitment includes self-selection (spotting the advertisement), followed by selection by the company. Biographical data, diagnostic interview, work samples and tests are the most important sources that can be used in the prediction of operator behaviour. We should like to comment on some of these sources.

The interview. This instrument is very often used by selecting committees, even for operator selection. However, there exists a lot of criticism about its reliability and validity for this purpose. In a review Guion (8) states: "...here and there are cases where some interviewers, in some situations, made judgements that yielded substantial coefficients of validity". The interview will always be used; neither the interviewers, nor the interviewees could imagine a selection process without it.

Process simulation as a work sample. During simulation research (Kragt (9)) it appeared that individual differences in performance occurred. These differences could partly be attributed to differences in personality traits: thoughtful, patience, or proneness to panicking. This operator behaviour and performance can be rated both objectively and subjectively. Such ratings form part of the Assessment Center Method (Daniëls et al. (10)). In the authors' view, such an approach is superior to psychological testing.
It appears in practice that the trial period of two months is used for further selection. Some of the interviewees were of the opinion however, that the real test took place later on during operator education, but especially in practice on the job as "qualified" operator!

Validity of operator selection procedures. Validation studies on operator selection are not known to the authors. But studies on fixing criterion-related validity for other functions are scarce too. In reference (4) it is stated: "...uncomfortable lack of published validation research of both objective and subjective selection techniques".
The predictive value of selection instruments, however, is greater as they more nearly resemble work samples (Reilley et al. (11)). That seems a good starting point for research on the use of process simulation in operator selection.

# EDUCATION & TRAINING

Operator education is defined as: all the activities aimed at learning - knowledge, skills and attitude - to obtain the overall capabilities needed for an operator job. Training is defined as: involving the operator in all the specific learning-activities (both theoretical and practical) needed for the operator in his function.

## Education & training needs

The content of operator education has been formulated from the more general needs which have been listed in the course of time by the companies. Thus the content reflects the policy with regard to the operator job. This policy appears to be: A process operator needs more than an accumulation of rules and skills, applicable only in a particular plant. It is a profession that requires background knowledge and understanding.

Training afterwards for particular plants should be based on training needs. Shepherd (12) advocates the employment of analysts who are able to trace and describe these needs. Annett & Duncan (13) developed a method by which the necessary operator activities can be described and analysed. In reference (12), however, it is stated that research conclusions on this topic are still scarce. An important reason why such methods have hardly ever been applied, is the time-consuming factor. The scarcity of task analysis for training, however, also occurs in other workplaces (Goldstein (14)).

## Operator education

In the Netherlands operator education is conducted by the Institute of "VAKOPLEIDING PROCESINDUSTRIE" (VAPRO); both theory and practice are involved. The theoretical part consists of general and professional education followed at a school. The practical part is provided by the company itself, according to the VAPRO directives. The VAPRO courses are legally accepted and can be considered as an all-round operator education, i.e. a basic level (knowledge, skill and attitude) for the operator profession. For a description of the contents of the courses we refer to the documentation (VAPRO, (15)).
A separation between the general professional education (VAPRO) and the plant-orientated training is approved of by both the interviewees and the authors. A separate professional education gives the operator professional status and contributes to the realization of the personnel policy that the authors would support.

## Job training

From all sides it is pointed out, that operator training has to be planned, beginning in the design phase of a plant and continuing as long as operators are working there (Wirstad et al. (16), (17); Verstoep et al. (18)). Moreover it appears, that companies are willing to discuss their training programs (Ratter et al. (19); Visser (20)).

Basic training. Theoretical knowledge, i.e. background knowledge and generally applicable rules about processes, pieces of apparatus and safe and efficient operation of plants concerned, is acquired in internal courses. These courses are given by the company's Training Department, mostly staffed by ex-production workers from the field. In several companies, motivated experienced operators are recruited to teach for several years. Practical knowledge & skill can only be acquired on the job ("learning by doing"). For the experienced operator to whom this task has been allocated, special demands are made with regard to technical knowledge, didactic skills and motivation.

Re-training and continued-training (reference (17)). Re-training is necessary for the operator to keep on functioning at an adequate level. The actual knowledge-level has to be revised. The exact knowledge, the right procedures have been forgotten, because they have been used only incidentally. Continued-training implies the updating from knowledge about changes in the plant itself; also the discussions about disturbances and breakdowns are part of this training. It takes place mostly on the job; too often, unfortunately, with use of log-book alone.
It seems rather difficult to realize both types of training in practice. The main reasons are:

- The operator cannot get time off from his job (leave or illness of his colleagues; this is an often voiced complaint, also by operators);
- The operator is not always motivated to follow further training ("what does it pay?");
- The lack of well-designed training programs; in each control room much

experience is present, but man-power is needed to transfer this experience into programs.

Simulation as a training method

As opposed to nuclear power stations and airplanes, in the chemical industry process simulation as a training tool is hardly used. The main reason is the high costs involved in realizing high fidelity simulation for a chemical plant. Nowadays, however, with distributed instrumentation off-line simulation for training is possible (Burks (21)).
Process simulation can be realized on different levels, each with its own fidelity. For instance, "to get the feeling of the process" needs another kind of simulation than "emergency-procedure training". Boeing (Seattle) gives pilots an off-line static procedure training whereas manoeuvres are taught in dynamic flight simulation. For a discussion about simulator validity, we refer to Moraal (22).
Finally, we should like to mention briefly "group simulation". Operators should be trained not only individually, but also as a group, because many operator tasks must be carried out in close co-operation. Their tasks are very interdependent. Working in small teams was seen as very important by all the interviewees. The possibilities for group-simulation games (Kuipers (23)) will have to be investigated, and also the techniques for the evaluation of education & training.

## JOB DESIGN

Design and re-design: a continuous process

Work organizations are open systems. Both technology and people are continually changing under the influence of the environment. There should be a continuous effort to match technology and people, and each to the environment. This is the area of job design: the organization of work, work situation and work conditions in such a way that the tasks fit the capabilities and interests of the workers as much as possible (Van Assen et al. (24)). We refer to the literature for the development of job design since the support for the socio-technical approach (Emery (25) and the works of Davis et al. (26), De Sitter (27) and Jenkins (28)).
In the following we deal with the design of the operator's job by using criteria for a meaningful job (Reference (24)): 1. semi-autonomous working groups; 2. accessibility of information; 3. job rotation; 4. work consultation; 5. job enrichment.

ad 1. Semi-autonomous working group. In process industries a team of operators is actually responsible for safe and economic process operation. Based on their research Blauner (29) and Vamplew (30) have stated: "...automated process technology demands highly integrated and co-ordinated workgroups to operate". Also Susman (31), Kuipers (32) and Bagchus et al. (33) regard the operator task as a typical group task. There is a constant demand for interdependence, co-operation and the ability to co-ordinate in this worksituation. In reference (29) it is found that such aspects explained why process operators preferred their job to their previous job in a non-process industry. In the classic study of Engelstad (34) the ideas of semi-autonomous working group were applied successfully.

ad 2. Accessibility of information. Operators have a central position in the information and communication network. Working on their knowledge about process operation, they can keep in touch and initiate discussions with others, higher in the hierarchy.

ad 3. Job rotation. One way of organizing team work is job rotation for the operators following a pre-programmed scheme. In practice it appears, that team work produces a greater flexibility of the operator; days-off, absenteeism and times of process breakdowns are then easy to cope with. Also task variety occurs. However, a disadvantage of this system is that sometimes operators have to move, because their capabilities do not fit the team structure. Another more realistic possibility is that some operators can perform all the tasks, whereas others can perform only some of them.

ad 4. Work consultation. In reference (24) work consultation is not only seen as an element of job design, but more particularly as a preceding phase in job design ("preventive job design"). It is quoted: "It is only common sense that a workplace can be best designed using the contributions of those who know most about the work..." (Job Reform in Sweden, (35)). Bailey (36) mentions a case in which the workers themselves (instead of their superiors or experts) suggested solutions for re-design.

ad 5. Job enrichment. Although Davis et al. (37) and reference (34) give examples to cases in which job enrichment has been applied to a great extent, the authors would have suggested a more limited use, because we are now thinking in terms of "multi-skilled" instead of "multi-crafted". Multi-skilled is used when job rotation occurs over several or all the specified jobs inside and outside one control room, or even over several control rooms, as already practised in some companies. Multi-crafted is used when something in addition to operator tasks, e.g. maintenance work, is also involved. In our view, however, the multi-crafted system is demanding too much of the operator.

Job design in the control room of tomorrow

It is surprising, that in the process industry the possibility was ever discussed at all of what is called "the separation of minds" in a control room. In reference (3) it is questioned whether it is efficient to choose an all-round operator education system. Would it not be better to split the education & training for control-room operator (panel operator) and plant operator? Ekkers et al. (38) remark that automation can lead to a greater separation between operators in the control room and plant operators. We support Lenior et al. (39) when they write against: "...a general belief that sophisticated instrumentation should be handled by sophisticated operators, while turning valves and starting pumps is more for rough-and-ready plant workers.". Our interviewees were of the opinion that such a separation should not be too rigorously pursued. Of course, both the function of panel operator and plant operator will be distinguished in the control room of tomorrow, but the authors agree with the interviewees that a total separation of panel operator and plant operator is wrong for the following reasons:

- The panel operator should know what happens behind the knobs and push-buttons. That he will only discover, when he himself is regularly carrying out the work outside the control room.
- He must experience what it means "to be sent out on a task". He will also have to realize what is the most efficient route.
- Specialism in the operator job is wrong. Each operator should be able to take a share in coping with a breakdown.
- The monotonous nature of the work (8 hours VDU work) demands a change of operator.
- In a small team together responsible for a safe and economic process operation, differences in status must be avoided as much as possible. In the old situation of reference (34) some operators felt "second grade" with all consequences that followed. Moreover, being a plant operator is not

inferior. Each operator of the team has to carry out this job in due time. On the one hand it is important for updating mental picture and mental model, on the other hand for reducing stress. Once a panel operator said to me: "...after a few days panel, I am really glad to move around in the plant.".

Limits of job design

Job design, as applied in the operator-process situation, could also mean the performance of maintenance and/or instrumentation work by the operator. But in many control rooms it would mean overload rather than challenge, because these tasks are of a different kind and require specific education & training that is different from the operator education & training. Moreover, in the authors' opinion, process control and process maintenance should be in different hands, for the sake of division of responsibility and competence. It is clear, of course, that the operator must have some ideas of maintenance, in order to know when to ask for help and how to assist maintenance personnel in the plant.

After reading the fore-going sections, it is not surprising that one is looking for alternative job designs in control rooms. Rijnsdorp (40) advocates an actual co-operation between Technical and "Social Designer"; by the latter he means: personnel staff, organizational experts, plant managers and representatives of the operators. Job design in which a better use is made of "human resources", and where there are improved, benefitted the organization. In reference (34) some interesting examples are mentioned.

## REFERENCES


1. Kragt, H. and Piso, E., 1983, Control-room vademecum; an ergonomic guideline (Deventer: Kluwer) (In Dutch).

2. Rijnsdorp, J.E. and Rouse, W.B., 1978, Design of man-machine interfaces in process control. In Proc. 5th IFAC/IFIP Internat. Conference on Digital Computer Application to Process Control (Edited by H.R. van Nauta Lemke and H.B. Verbruggen) (Amsterdam: North Holland), pp. 705-720.

3. Boot, J.C.D., 1979, Changes in the operator's job. Paper presented at the joint meeting of the VAPRO board of directors and its consultancy board, 24th August. (In Dutch.)

4. Tenopyr, M.L. and Oelten, P.D., 1982, Personnel selection and classification, Annual Review of Psychology, 33, pp. 581-618.

5. Hiscock, W.G., 1938, Selection tests for chemical process workers. Occupational Psychology, 12, pp. 178-186.

6. Brenninkmeijer, G., 1964, Working in Automated Factories (Amsterdam: Swets and Zeitlinger). (In Dutch.)

7. Lees, F.P., 1980, Loss Prevention in the Process Industries (London: Butterworths).

8. Guion, R.M., 1976, Recruiting, selection and job placement. In Handbook Industrial and Organizational Psychology (Edited by M.D. Dunnette) (Chicago: Rand McNally College Publishing Company).

9. Kragt, H., 1984, A comparative simulation study of annunciator systems. This symposium. (Also accepted for publication in Ergonomics.)

10. Daniëls, M.J.M. and Roe, R.A., 1983, First Principles of Personnel Appraisal. (In Dutch; in press.)

11. Reilley, R.R. and Manese, W.R., 1979, The validation of a minicourse for telephone company switching technicians, Personnel Psychology, 32, pp. 83-90.

12. Shepherd, A., 1983, Training process operators. Paper presented during the joint Symposium on Human Reliability in the Process Control Centre, Manchester 20/21 April, ICE, IM&C and SRS.

13. Annett, J., Duncan, K.D., Stammers, R.B. and Gray, M.J., 1971, Task Analysis (London: H.M.S.O.).

14. Goldstein, I.L., 1980, Training in work organizations. Annual Review of Psychology, 31, 229-272.

15. VAPRO, 1982, Complete Programs for Process Operator A and B ('s Gravenhage: Training Board for Process Industries). (In Dutch.)

16. Wirstad, J. and Andersson, H., 1980, Competency and training requirements for operators. In Automation for Safety in Shipping and Offshore Petroleum Operations, IFIP/IFAC Symposium, Trondheim Norway (Editors: A.B. Aune and J. Vlietstra) (Amsterdam: North-Holland Publishing Company), pp. 225-229.

17. Wirstad, J., 1983, Development of operator training as an integrated part of the specification, design and implementation of a process control system. In Training for Tomorrow, preprints IFAC/IFIC Conference Leiden, The Netherlands (to be edited by: J.E. Rijnsdorp and Tj. Plomp) (Oxford: Pergamon Press Ltd.), pp. 13-21.

18. Verstoep, N.D.L. and Bessant, J., 1983, Making the future: developing the individual and the organization. In Training for Tomorrow (see Wirstad, 1983), pp. 204-212.

19. Ratter, H. and Plant, B.M., 1983, Benefits resulting from structured training of process operators on multi-product chemical site. Paper presented during the joint Symposium on Human Reliability in the Process Control Centre, Manchester 20/21 April, ICE, IM&C and SRS.

20. Visser, A.D., 1983, Training in the process industry. In Training for Tomorrow (see Wirstad, 1983), pp. 78-84.

21. Burks, H.C., 1981, Training aspects of multi-variable advanced control systems. In preprints IFIP Third World Conference on Computer Education (Edited by R. Lewis and E.D. Tagg) (Amsterdam: North-Holland Publicity Company).

22. Moraal, J., 1981, Evaluating Simulator Validity. In Manned System Design (Edited by J. Moraal and K.F. Kraiss) (New York: Plenum Press), pp. 411-426.

23. Kuipers, H., 1980, From Competition to Co-operation; a simulation experiment as a technique for changing work attitudes. Ph.D. Thesis, Leiden University. (In Dutch.)

24. Assen, A. van and Hertog, F. den, 1981, Job design; from task rotation to organizational design. In Handbook Work and Organizational Psychology. - afl. 3; 4.8Ass.1 - 4.8Ass.34 (Edited by P.J.D. Drenth, H. Thierry, P.J. Willems and Ch.J. de Wolff) (Deventer: Van Loghum Slaterus). (In Dutch.)

25. Emery, F.E., 1959, Characteristics of socio-technical systems. In Design of Jobs (see Davis et al., 1972).

26. Davis, L.E. and Taylor, J.C., 1972, Design of Jobs (Harmondsworth: Penguin Books Ltd.).

27. Sitter, L.U. de, 1981, On the Road to New Factories and Offices (Deventer: Kluwer). (In Dutch.)

28. Jenkins, D., 1981, QWL: - Current trends and directions. Issues in the Quality of Working Life; a series of occasional papers, no. 3, december (Ontario: Quality of Working Life Centre).

29. Blauner, R., 1964, Alienation and Freedom (Chicago: The University of Chicago Press).

30. Vamplew, C., 1973, Automated process operators: work attitudes and behaviour. British Journal of Industrial Relations, 11, pp. 415-431.

31. Susman, G.I., 1972, The impact of automation on work group autonomy and task specialization. In Design of Jobs (see Davis et al., 1972).

32. Kuipers, H., 1973, Aspects of leadership in the process industry, Mens en Onderneming, 27, pp. 94-105. (In Dutch.)

33. Bagchus, P.M. and Kuipers, H., 1982, Taskgroups in Organizations, M & O, 36, pp. 481-497. (In Dutch.)

34. Engelstad, P.H., 1974, Socio-technical approach to problems of process control. In The Human Operator in Process Control (Edited by E. Edwards and F.P. Lees) (London: Taylor & Francis Ltd.).

35. Job Reform in Sweden, 1975 (Stockholm: SAF).

36. Bailey, J., 1983, Job Design and Work Organization (London: Prentice-Hall International).

37. Davis, L.E. and Sullivan, C.S., 1980, A labour-management contract and quality of working life. Journal of Occupational Behaviour, 1, pp. 29-41.

38. Ekkers, C.L., Brouwers, A.A.F., Pasmooij, C.K. and Vlaming, P.M., de, 1980, Human Control Tasks (Leiden: NIPG/TNO). (In Dutch.)

39. Lenior, T.M.J., Rijnsdorp, J.E. and Verhagen, L.H.J.M., 1980, From field operators to central control room operators, Ergonomics, 23, pp. 741-794.

40. Rijnsdorp, J.E., 1982, State-of-the-art and future trends in the field of man-machine systems. In Analysis, Design, and Evaluation of Man-Machine Systems, IFAC Conference Baden-Baden (to be edited by: G. Johannsen and J.E. Rijnsdorp) (Düsseldorf: VDI), conclusions.

# JOB DESIGN IN COLLIERY CONTROL ROOMS

R. Martin, G.C. Simpson & C.F. Talbot*

The paper is concerned with the assessment of the effects of new technology on workers' perceptions of their jobs in colliery control rooms. The study described attempts to identify areas of the control room operator's job, from questionnaire results, where any change in job design and organisation may improve job satisfaction, motivation and reliable performance.

## INTRODUCTION

For the past 30 years, there has been a growing interest in the human factors aspect of operator performance in control rooms. Initially, during the 50's and 60's, this work concentrated on the direct interaction between the operator and individual task elements such as dials, controls etc. (e.g. Spencer (1)). As technology developed and control systems became more integrated, the interests of human factors widened to include not only the adaptation of the early studies to new technology interfaces such as VDUs, but also a development of more detailed studies aimed at understanding the operators' mental processes and decision skills (e.g. Bainbridge (2), Brigham and Lasio (3)). More recently, as control technology has developed further, the emphasis has shifted towards the study of the quantification of human error in increasingly automated plant operations.

Running parallel to these developments in "hard" ergonomics, in other industrial situations, has been an increasing concern about the influence of psycho-social factors (e.g. job satisfaction, organisational and supervisory influences, stress and general well-being etc.) on performance (e.g. Broadbent and Gath (4), Buchanan and Boddy (5)).

These two streams of development can be combined to present a schematic representation (Figure 1) of the human factors issues which need consideration in the evaluation and design of effective new technology man:machine systems.

* Ergonomics Branch, Institute of Occupational Medicine, c/o National Coal Board, Mining Research and Development Establishment, Ashby Road, Stanhope Bretby, Burton upon Trent, Staffordshire, DE15 OQD

In order to ensure the most effective, safe and efficient operation of new systems, the factors shown in Figure 1 must be considered as an integrated whole. Considerable work has been carried out in relation to the central area: hardware design and layout (summarised in, for example, Cakir et al (6) and Grandjean and Vigliani (7)) and Best et al (8) have described ways of improving the dissemination to, and use of such information by design engineers. The second layer is a topic of considerable current research interest, (e.g. Kemp et al (9), Hebditch (10)) however, the organisational and job design aspects have received relatively little attention, especially in complex control room situations.

This study is concerned, therefore, with the assessment of the effects of new technology on workers' perceptions of their jobs in colliery control rooms.

Computer-based control and monitoring systems are being installed in colliery control rooms throughout the British mining industry. Systems are being introduced which cover the monitoring and control of underground coal clearance systems, the monitoring of mine environment and the monitoring of production plant status. For such primary systems, computer-based management information systems are also being developed. Although the introduction of such systems inevitably changes the nature of operators' jobs, in general the job of the control room operator whether computer-assisted operations or "traditional" control rooms remain the same. In essence, the colliery control room is the information relay centre between underground and surface involving a mixture of routine operations, with predictably busy periods of shift change with the constant possibility of emergency situations which can have major safety or production implications.

Ideally, studies of the impact of new technology on operator perceptions and job design should be carried out longitudinally with studies on the same group before and after change. Unfortunately, this was not possible in this project. A cross-sectional approach had to be utilised comparing operators in collieries with traditional control rooms against those in computer-aided control rooms. As far as possible, the collieries were matched on the basis of parameters associated with the size and complexity of the colliery operation and organisation.

Given the problems inherent in any investigation carried out during normal operations, it was felt that only a single questionnaire investigating job design in colliery control rooms could be used. After surveying the literature, the Job Diagnostic Survey (Hackman and Oldham (11)) was chosen on the basis of its comprehensive underlying model and its successful use in industrial situations. A brief description of the technique is given below.

## THE JOB DESIGN STUDY

### The Job Characteristics Model

Figure 2 represents the model proposed by Hackman and Oldham (11) to interpret the Job Diagnostic Survey (JDS). The Job Characteristics Model proposes that affective outcomes of high internal work motivation, high work satisfaction, high quality performance and low absenteeism and turnover came about when three critical psychological states are present. These critical psychological states are the experienced meaningfulness of the work, the experienced responsibility for the outcomes of the work, and the knowledge of results of the work activities. Three core job characteristics, those of skill variety, task identity, and task significance are especially powerful in influencing the experienced meaningfulness of work.

The characteristic of the job which increases the feelings of personal responsibility for the work outcomes is automony, and the knowledge of results of work is directly affected by the amount of feedback from the work.

Through the critical psychological states, the affective outcomes are fostered and enhanced. The core job characteristics are said to enhance these psychological states and through them foster the affective outcomes.

Although the affective outcome variables are basically fostered by the core job characteristics, there are many attributes of the individual which effect how they respond to their work. The JDS incorporates three of these which are termed 'moderators' in the model. These are the job relevant knowledge and skill, the individual growth need strength, and the 'context' satisfactions.

In other words, these represent the extent to which the operator perceives himself as having the required knowledge and skill, his needs to develop beyond the boundaries of the existing job and his satisfaction with the 'context' factors such as pay, working conditions etc..

The general results are discussed below in terms of the main components of the model (full details of the analysis are given in Best et al (12)).

## RESULTS

### Main Components

The affective outcomes (internal work motivation, general satisfaction and growth satisfaction) were perceived differently by operators in computerised and traditional control rooms. The results show that the traditional control room operators scored higher on each of the affective outcomes. However, only two of the differences were statistically significant: internal work motivation and growth satisfaction. A comparison of mean scores is shown in Figure 3.

On the basis of this study, therefore, the operators in traditional control rooms were generally more satisfied with their job than their counterparts in computerised control rooms. To examine the influential factors, it is necessary to consider other aspects of the model.

### Core Job Characteristics

The response of the two groups on the five core job characteristics are shown diagrammatically in Figure 4. It can be seen from Figure 4 that there were few real differences between the scores obtained from computerised and traditional control rooms. It is, however, worth noting that although differences were slight, operators in traditional control rooms were consistently more positive in their perceptions. This is particularly surprising in relation to skill variety and task significance (which was significantly different) where there are reasonable *a priori* grounds on which to expect higher scores in the computerised control rooms.

The Hackman and Oldham (11) model also utilises a combination of the job characteristic scores to form the Motivating Potential Score (MPS). It is postulated that this score reflects the extent to which the intrinsic job components create a motivating potential, i.e. it shows the extent to which operators are likely to find the job itself as motivating. The MPS was significantly higher for traditional control room operators than those in computerised control rooms. However, in terms of the model's predictions neither score could be considered as representing a low MPS. Perhaps the most useful interpretation of this result is that computer-aiding in the control room has clearly not improved the motivating potential of the job.

### Context Satisfaction

The context satisfaction parameters measured by the JDS are job security, pay, satisfaction with co-workers and supervision.

No differences between operators' perceptions of pay and job security were recorded. This is consistent with the reality of the situation in that payment grades are common to both and within the sample there were no obvious "threats" to job security in any of the collieries. However, operators in traditional control rooms were significantly more satisfied with co-workers and supervision.

### Growth Need Strength

The comparison of growth need strength (a measure of the operator's need for personal accomplishment at work) between traditional and computerised collieries showed no significant differences.

## DISCUSSION

The complexity of the Hackman and Oldham (11) model has meant that a full description of the results and analytical interpretation has not been possible in such a short paper. However, one analytical interpretation shown to be of value in previous studies (e.g. Oldham (13), Oldham et al (14)) is particularly interesting in relation to the present results. This posulates that the final affective outcome scores (the overall descriptors of satisfaction) are a function of the relationship between the motivating potential score, the growth need strength and the context satisfactions. In the present study, no significant differences were found in relation to growth need strength, however, significant differences were obtained on aspects of the motivating potential score and context satisfactions. It would appear plausible, therefore, that the reduced growth satisfaction and internal work motivation in computerised control rooms, as indicated in the affective outcome scores, is a function primarily of the reduced satisfaction with co-workers and supervision. They must also be a partial influence of the significantly reduced perception of task significance and the resultant lower motivating potential score.

This clearly emphasises the importance of organisational factors in ensuring the best utilisation of new technology. However, this result may not be exclusively organisational. In parallel with the JDS studies, the project investigated the "acceptability" of management information in computerised and traditional control rooms. In this study there were frequent perceived limitations in the information presented in both traditional and computer-based systems but, perhaps surprisingly, more so in the computer-based systems. It is possible, therefore, that the reduced satisfaction with co-workers and supervision could reflect, in part, a perceived reduction in the "quality of service" provided by the control room.

While this speculation is plausible from results of the two studies and consistent with the highly interactive nature of the problem of new technology as described in Figure 1. It should be emphasised that the JDS and management information studies in this project were carried out independently. Neither can the co-worker and supervision satisfactions identified in computerised control rooms be attributed to any change in control room operator personnel. Worker selection is similar in both types of control room.

The interpretation of these results is highly complex, however, it emphasises the impact new technology can have on many other functions associated with the job. A thorough investigation of these considerations must be carried out if positive change is to result. It must remain, therefore, at this stage a topic for further study.

## REFERENCES


1. SPENCER,J.,1968,The ergonomics of panel layout,Measurement and Control,1,145.

2. BAINBRIDGE,L.,1971,The influence of display type on decision making,IEE Conference "Displays",Loughborough University,Reprinted in IEE Conference Publication,80.

3. BRIGHAM,F.R.,and LAIOS,L.,1975,Operator performance of a laboratory process plant,Ergonomics,18,1.

4. BROADBENT,D.E.,and GATH,D.,1979,Chronic effects of repetitive and non-repetitive work,Ergonomics Society Conference,"Psychophysiological Response to Occupational Stress",Reprinted in Mackay,C.,and Cox,T., (Eds),"Response to stress : Occupational aspects",Guildford : IPC.

5. BUCHANAN,D.A.,and BODDY,D.,1983,Advanced technology and the quality of work life : The effects of computerised controls on biscuit making operators,Journal of Occupational Psychology,56,2.

6. CAKIR,A.,HART,D.J.,and STEWART,T.F.M.,1979,The VDT Manual,Darmstadt, Inca-Fiej Research Association.

7. GRANDJEAN,E.,and VIGLIANI,E.,1980,Ergonomics aspects of visual display terminals,London : Taylor and Francis.

8. BEST,C.F.,SIMPSON,G.C.,and MASON,S.,1983,Ergonomics in control room design,building from basics,Paper presented at Joint Institute of Chemical Engineers, Safety & Reliability Society Institute Measurement and Control Conference,"Human Reliability in the Control Centre", Manchester.

9. KEMP, S.,GOTHELF,R.,CAREY,M.,TAYLOR,R.G.,and WHITFIELD,D.,1983, Ergonomic aspects of menu based dialogues for man-computer interaction,Paper presented at Ergonomics Society Annual Conference, Cambridge.

10. HEBDITCH,D.,1979,Design of dialogues for interactive commerical applications,Infotech States of the Art Conference,"Man/Computer Communication",Reprinted in "Man/Computer Communication,2, Maidenhead : Infotect International Limited.

11. HACKMAN,J.R.,and OLDHAM,G.R.,1980,Work Redesign,Addison-Wesley.

12. BEST,C.F.,FERGUSON,C.A.,MARTIN,R.,MASON,S.,SIMPSON,G.C.,and TALBOT,C.F.,1984,The human aspects of computer-based monitoring and control of mining operations,Final Report on CEC Contract 7206/00/8/01,Edinburgh : Institute of Occupational Medicine (in preparation).

13. OLDHAM,G.R.,1976,Job characteristics and internal motivation : The moderating effect of interpersonal and individual variables,Human Relations,9(6),559-569.

14. OLDHAM,G.R.,HACKMAN,J.R.,and PEARCE,J.L.,1976,Conditions under which employees respond positively to enriched work,Journal of Applied Psychology,61,395-403.

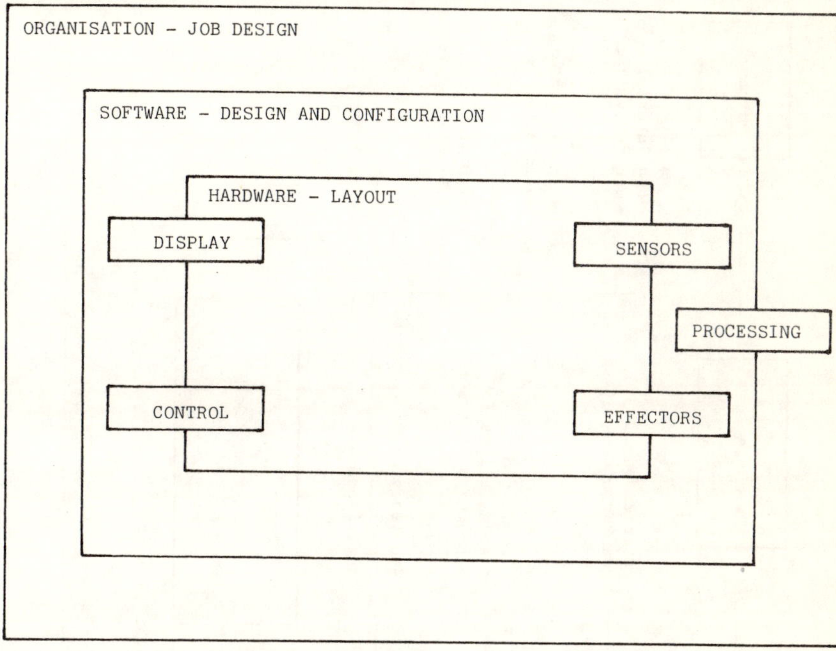


FIGURE 1 Schematic Diagram of the Human Factors Considerations in New Technology Man:Machine System Design

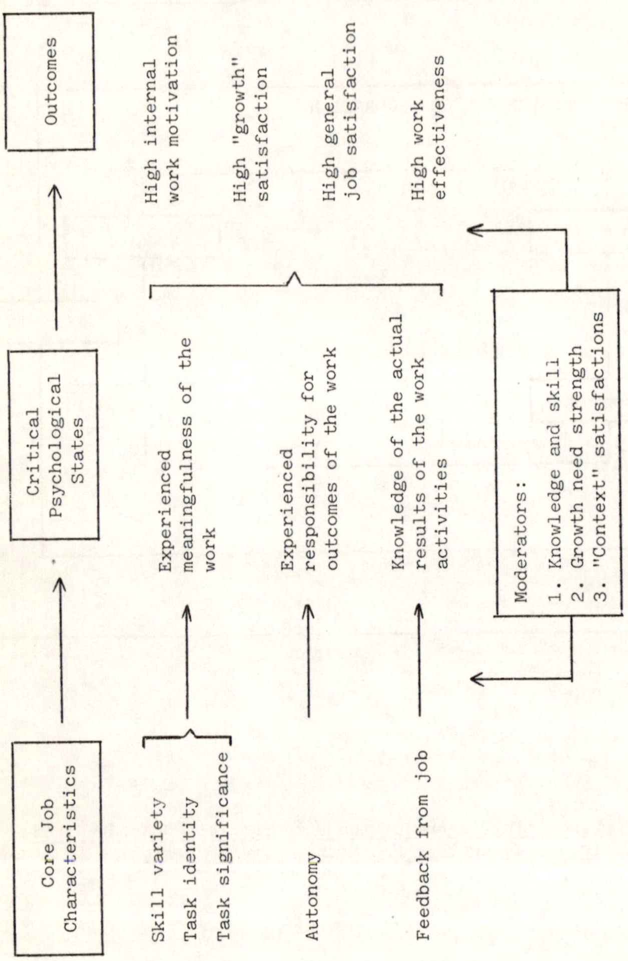


FIGURE 2 The Complete Job Characteristics Model

FIGURE 3 Mean Affective Outcome Scores from Operators in Traditional and Computerised Control Rooms

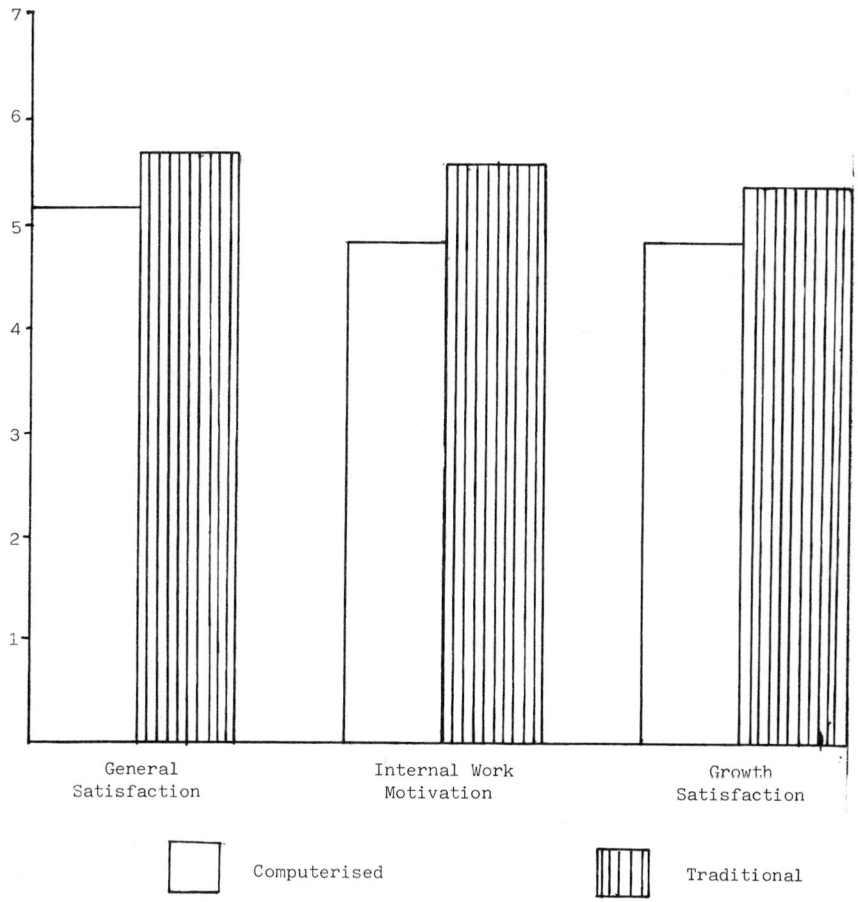

FIGURE 4 Mean Core Job Characteristic Scores in Traditional and Computerised Control Rooms

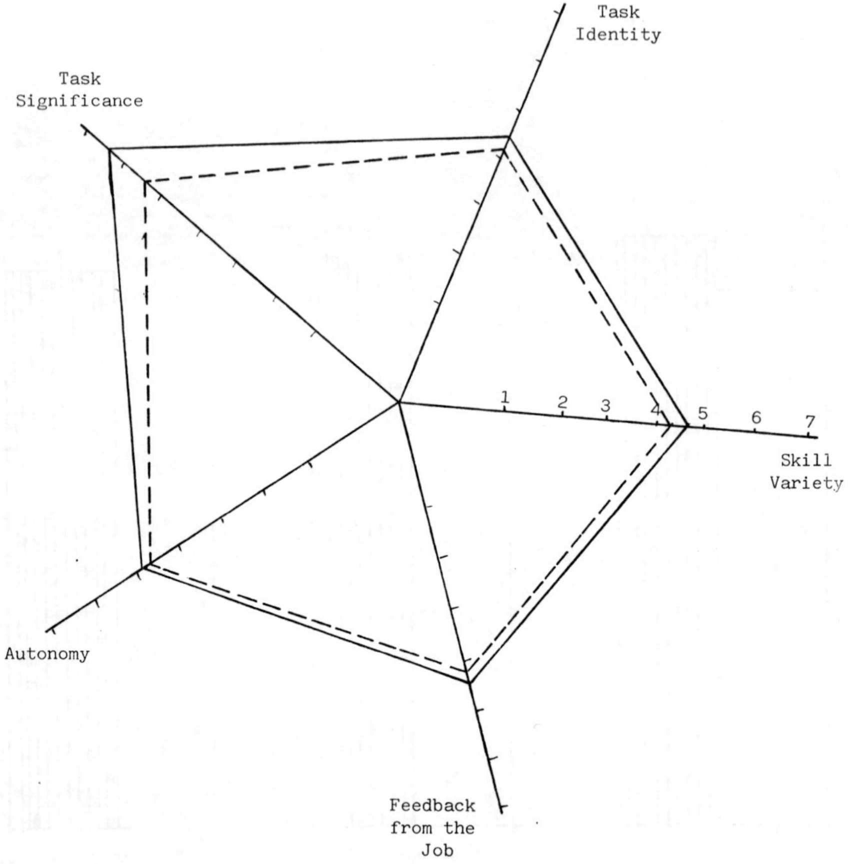


Key:

------------ Computerised ———————— Traditional

# STUDIES IN DEVELOPMENT OF METHOD TO ASSESS THE MANAGEMENT FACTOR IN INDUSTRIAL LOSS

J. Powell and D. Canter *

Hazardous occurrences in the chemical industry are a major source of loss. Such incidents are more often caused by deficiencies in management systems and other human-related aspects than by technical failures in hardware systems. Human factors can also have a great effect in preventing escalation to a major loss. There is no currently available technique for identifying and evaluating the influence of such factors on the losses associated with the operation of chemical processes. The development of a preliminary model for this purpose will be described, and its future potential applications discussed.

## BACKGROUND

If we are to reap the benefits of increased scale and complexity of technology, we need to ensure that the advantages it offers are not negated by poor associated systems of management. Unfortunately, the evidence from accident case histories such as Flixborough (Parker (1)), Seveso (Temple (2)), and Three Mile Island (Malone et al (3)), suggests that there is much room for improvement in the human and social component of industrial organisations. There have been advances in understanding and producing techniques for the evaluation of the contribution to losses made by the hardware components of systems (eg. Hazards and Operability Studies, Hazards Analysis, Probabilistic Risk Assessment among others) but, as noted by Cox and Comer (4), comparatively little in understanding the important human and social elements. The most relevant development in the latter field has been the production of techniques for assessing human reliability in process control, see for example, Embrey (5). Such methods, however, are not designed to take into account the entire range of human-related factors which influence the control of hazards and the ability to cope with their consequences, but are more appropriate for analysing more specific elements of task performance.

For the past two years, the Department of Psychology at the University of Surrey has been involved in a research project designed to produce a workable tool for the evaluation of the contribution made by human and social factors to losses arising through fires and explosions in the chemical and petro-chemical industry. This work has been initiated and sponsored by the Insurance Technical Bureau (London) which has long recognised the potential value of such a technique (Singh and Munday (6)), and supported by International Oil Insurers. The department's own interest in this field

* Department of Psychology, University of Surrey.

stems from its involvement over the past seven years in a project which has investigated human behaviour in fires, as described by Canter (7). There are obvious parallels between this research and studies of controlling and coping with hazardous situations in chemical plants, which also involves consideration of such factors as the efficiency of training, communications and emergency alarm systems.

## THE MANAGEMENT FACTOR IN THE CONTEXT OF THE INSTANTANEOUS FRACTIONAL ANNUAL LOSS ALGORITHM

The Management Factor technique has initially been developed in the context of the Instantaneous Fractional Annual Loss (IFAL) algorithm, as described in reference (6), which provides a way of assessing expected average annual financial losses arising through fires and explosions in chemical and petrochemical processes expressed as a fraction of the total at risk. IFAL has three components, the process factor (p), the engineering factor (e) and the management factor (m) so that:

$$\text{IFAL} = p.e.m \qquad (1)$$

where 'p' represents the inherent hazard of the process
'e' represents standards of engineering design and construction
'm' represents the quality of operational management

The IFAL value for a process is numerically equal to p when the plant is designed, constructed and managed according to Standard Good Practice (SGP), i.e. when m=1. Levels of the e and m factors which are better or worse than SGP will modify the IFAL value accordingly. Managerial and operational procedures deemed to be in accordance with SGP for a particular process will not necessarily be in accordance with SGP for another process. This is because what constitutes SGP depends on the processes used. Therefore, identical managerial and operational procedures applied to different plants may result in different values for the Management Factors associated with different plants. In other words, the Management Factor needs to be 'process specific'. The technique also has to be suitable for use by engineers and assessors who may not have expertise in social science. This means that the method must be logical, easily understandable, and simple to use in the practical situation.

## PRELIMINARY WORK ON THE MANAGEMENT FACTOR TECHNIQUE

### Identifying What to Measure

At the beginning of this study, an attempt was made to take stock of the information relevant to the future development of such a technique. The major sources available to us were the findings from social psychological and related research fields, accident case histories, and the literature pertaining to industrial risk assessment. The first aim was to produce a comprehensive list of human and social factors which may influence the occurrence of fires and explosions and their associated consequences. In the light of available knowledge from these different areas, a list was drawn up of about 50 human and social related items which are known to influence loss. These were referred to as the 'Management Factor Items' and could be classified under the following general headings:

1. Organisational Structure - including such items as identifiability of roles, accountability for responsibilities related to loss control and the efficiency of plant communications.

2. Risk Management - including consideration of the formality of procedures for identifying, evaluating and treating risk, and the communication of risk to personnel.

3. Manning - aspects of training, selection, satisfaction with the social and physical working environment.

4. Emergency Measures - including emergency training,testing of fire, gas and other alarm systems, recording of incidents and learning from past mistakes.

5. General Management - including working procedures, permit-to-work systems, safety audits and/or inspections.

At this stage we were rather naive about the internal organisation and functioning of process plants and decided to attempt to clarify our initial ideas by visiting a number of plants and discussing the proposed project with individuals who have experience in plant management, plant operation, safety management or loss control. These interviews were very informal in the sense that individuals were allowed to freely express their own views about which Management Factor Items are important to loss control, but were structured to some extent through the use of our preliminary list. This exercise was very valuable in terms of clarifying our ideas and enabling the production of a revised list of 46 Management Factor Items. There was a high degree of consensus in the expressed opinions: for example, 'good standards of plant housekeeping' and 'clear working conditions for contractual workers' were stated as being of great importance by all respondents.

## Introduction of Loss Control Components

A very simple way of evaluating the Management Factor from the list of the 46 Management Factor Items would be to visit a plant, assess the standard for each item on a common scale, and use the mean score as the Management Factor Index. This method, however, would be insufficiently informative in the sense that it would not be structured according to any relationships between the Management Factor Items, or to their relative importance in terms of their potential contribution to loss. In order to impose more structure on the process, it was decided to introduce the concept of 'Loss Prevention Components' from the IFAL 'p' factor method for evaluating inherent process hazards through fires and explosions. Table 1 illustrates how 'Loss' can be broken down into identifiable components which bear direct relevance to the problem at hand.

## Relating the Management Factor Items to Components of Loss

Clarification of Ideas through Expert Rating Exercise. A subsequent development from the introduction of these components was the realisation that there would now be a firmer basis for determining the relative importance of individual Management Factor Items to the overall Management Factor. This could be achieved by considering the Management Factor to be composed of the contributions made by the 4 Primary Loss Prevention Objectives (CP's) so that:

TABLE 1 - Components of Loss through Fires and Explosions.

| Primary Objectives | Subobjectives |
|---|---|
| Control of Loss of Containment | Control of internal explosions<br>Control of corrosion and erosion<br>Control of machine failure<br>Control of operator error<br>Control of collisions with equipment |
| Control of Ignition | Control of background ignition sources<br>Control of naked flame sources |
| Control of Spread | Rapid detection of material release and fire or explosion<br>Rapid assessment of material release and fire or explosion<br>Correct and efficient action after material release and fire or explosion |
| Control of Damage | Good standards of housekeeping |

$$\text{'m' (the Management Factor)} = CP_1 \cdot CP_2 \cdot CP_3 \cdot CP_4 \quad \ldots\ldots\ldots\ldots(2)$$

In turn, each CP would be composed of a number of Loss Prevention Subobjectives. The relative importance of each Management Factor Item could be incorporated into the model by a fixed set of weighting coefficients for each Management Factor Item against each Loss Prevention Subobjective. In order to attempt to obtain a preliminary set of such weightings we again enlisted the help of a number of experts and asked them to rate, on a scale from 1 to 9, the relative contribution made by each of the 46 Management Factor Items listed in Table 2 to each of the Loss Prevention Subobjectives in Table 1. The incorporation of expert judgement has been an important feature in the development of the Management Factor Model. It represents a deleibreate attempt to avoid dependence on retrospective data drawn from past incidences, which, by their very nature are too rare to justify subsequent extrapolation. The use of expert opinion is gaining favour in a number of related fields: see for example reference (5).

Six individuals with extensive experience of process plant operation and/or of plant safety assessment participated in the exercise. They were given a list of the 46 Management Factor Items and asked to record their

TABLE 2 - List of Management Factor Items.

1. A positive committment by management to maintain the highest standards of safety.
2. A clear line of accountability for safety with individual responsibilities clearly defined.
3. Regular meetings of safety committees which involve personnel at different levels in the company hierarchy, have a useful content, and the outcome of which is well-documented.
4. Good patterns of communications between all those individuals who need to exchange information about matters related to loss prevention.
5. A well-written safety policy which is up-to-date and readily available to anyone who may wish to consult it.
6. Regular safety audits or inspections which are thorough and comprehensive and the results of which are well-documented.
7. Frequent and regular tests of fire and gas alarms.
8. Planned emergency practices carried out at regular intervals and the outcome of which is noted as a basis for improving existing procedures.
9. The training of personnel in basic fire-fighting techniques.
10. Liaison with the local fire brigade regarding actions to be taken in the case of a major emergency.
11. The existence of well-marked and fully-functional fire extinguishers and manual call points.
12. Clear access to emergency exits and access routes at all times.
13. Clear access to fire extinguishers and other emergency equipment at all times.
14. The existence of a well trained and competent maintenance force.
15. The ability to hear fire and gas alarms throughout the plant.
16. The existence of, and adherence to clearly defined and documented procedures for making changes to any past of the process or plant.
17. Comprehensive maintenance records.
18. Specialist maintenance engineers to back up plant maintenance teams.
19. Recording of all equipment failures.
20. The adherence to clearly documented permit-to-work systems for any difficult or dangerous work.
21. The existence of, and adherence to clearly defined conditions of work for ouside contractors.
22. Clear process operating instructions which are revised as soon as it is necessary.
23. The documentation and use of data relating to previous 'near misses' and minor incidents within the plant.
24. Close proximity of the plant manager's office to the process site in order that he can make frequent visits to monitor and check conditions.
25. The existence of safety incentive schemes to encourage safe working.
26. The provision of safety showers and eye-wash bottles/ baths.
27. The provision and wearing of all forms of protective clothing which are needed for safe working.
28. Adherence to any special Health and Safety legislation which applies.
29. Good standards of housekkeping in the plant.
30. Current financial stability of the organisation which operates the process.
31. A clear basis on which to allocate money for different areas related to loss prevention associated with the process.
32. Financial priority for areas related to loss prevention.

/continued over

ctd.

33. Formalised and documented procedures for identifying and assessing the potential risks of hazards associated with new or existing processes.
34. Continued attempts to reduce the known risks existing in the process.
35. An absence of any previous major fires and explosions.
36. Low rates of lost time accidents.
37. Low rates of absenteeism and labour turnover throughout the plant.
38. Competent and well trained process and control operators.
39. Competent and well trained management.
40. A workforce which is satisfied with its working conditions.
41. Sufficient cover for absenteeism through holidays and illness.
42. A safety conscious workforce.
43. An appreciation by management of the need for controls and equipment which are well designed from the human factors point of view.
44. The involvement of human factors specialists in plant design.
45. Comfortable thermal, aural and visual working environments in all areas of the plant.
46. A sociable working environment.

---

numerical ratings on a 46 x 11 response grid. The judges completed the exercise individually and were allowed as much time as they needed. They were encouraged to check their responses and revise initial decisions where appropriate before returning the completed response sheet.
The overall aims of the assessment exercise were:

1. To clarify the structure of the Management Factor Items.
2. To clarify the structure of the Loss Prevention Subobjectives.
3. To investigate whether such an exercise could be used to derive the coefficients required to rate each Management Factor Item against each Loss Prevention Subobjective.

A Multidimensional Scaling technique called Smallest Space Analysis (SSA) as described by Romney et al (8) was used to determine any structures existing in the Management Factor Items in terms of the manner in which they were rated overall with respect to the Loss Prevention Subobjectives. The output of the SSA program provides the clearest illustration of the way in which the Management Factor Items were rated in terms of their relative importance to the Subobjectives. The output provides plots of the 46 Management Factor Items in each of 4 dimensions where distance between items in the plots is inversely proportional to the degree of positive correlation between those items. On examining these plots, it is possible to partition the space containing the items in such a way that items with some quality in common are enclosed in the same region. The outcome of this for the data from the weighting exercise is shown in Figure 1. A careful examination of the qualities of the items in each group enabled them to be classified in two ways: firstly according to one of six 'aspects' of loss control activity and secondly to a 'context' of either 'State of the Organisation' or 'Working Procedures'.

| Group 1 'Aspects' | Group 2 'Contexts' |
|---|---|
| Executive safety | State of the organisation |
| General safety | Working Procedures |
| Preventative measures | |
| Emergencies | |
| Documentation | |
| Human/social factors | |

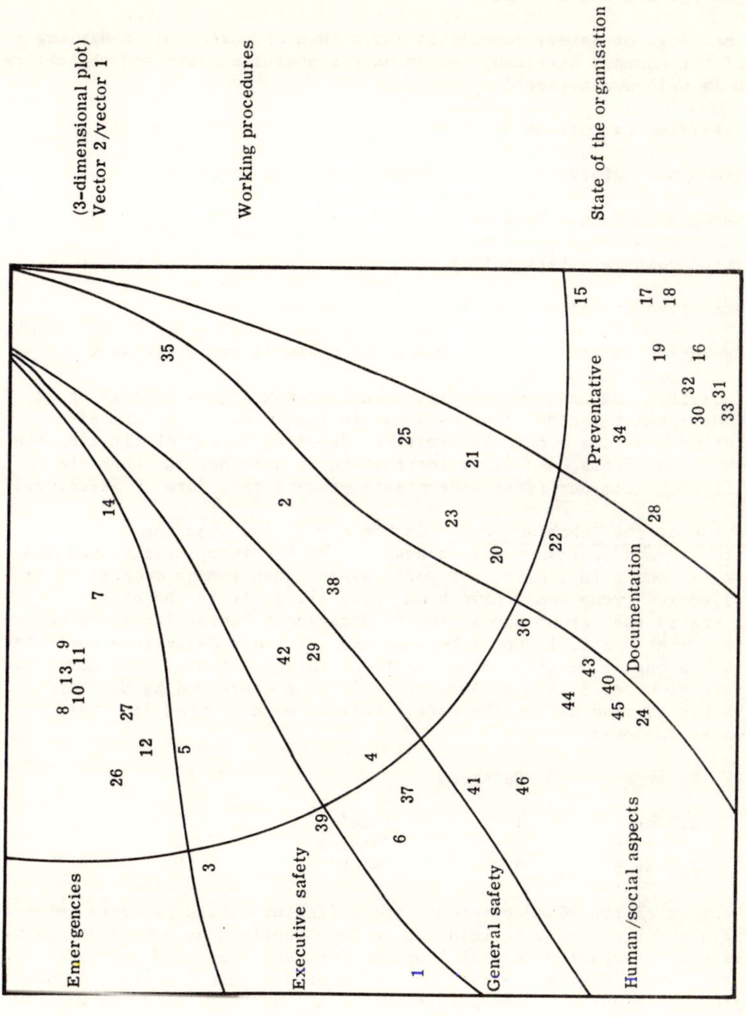


Figure 1: 46 Management Factor Items Grouped on Smallest Space Analysis Plot.

Each of the Management Factor Items can thus be classified in two ways. For example, Management Factor Item 3:

'Regular meetings of safety committees which involve personnel at different levels of the company hierarchy, which have a useful content and the outcome of which is well-documented'

can be classified as follows:

Aspect: Executive safety Context: Working Procedures

whereas Management Factor Item 5:

'Well written compnay safety policy'

can be classified as follows:

Aspect: Executive safety Context: State of the Organisation

All possible combinations of aspects and contexts give a total of 12 groups of Management Factor Items as shown in Figure 1. This form of classification provides a more informative, descriptive way of treating the Management Factor Items, and it is interesting to note how the emphasis is rather different from our first attempts to produce some form of structure.

Clarification of the Subobjectives. On the assumption that the 11 Subobjectives are a representative subset of the 4 Primary Loss Prevention Objectives, as shown in Table 1, we would expect that common members of each primary objective group would have been rated similarly to the other subobjectives in that group across the 46 Management Factor Items. Also, we would expect that the differences between the 4 Primary Objectives would be represented through some differences in their rating patterns. To investigate this, a Partial Order Scalogram Analysis (POSA) as described by Shye and Elizur (9) was carried out on the data. Ratings were grouped into three categories as follows:

| Ratings | Category | |
|---|---|---|
| 1,2 or 3 | 1 | 'Low' |
| 4,5 or 6 | 2 | 'Medium' |
| 7,8 or 9 | 3 | 'High' |

The output of the POSA represents the different rating patterns (profiles) for Management Factor Items in relation to Subobjectives as points in space, and individual ratings for the 11 Subobjectives are presented in the form of 11 separate plots. Figure 2 shows the POSA plot for subobjective 2 'Control of corrosion and erosion'. The partitions group the items into their three different rating categories. Each item is represented by a point in space and this configuration is identical for all 11 subobjective plots. The difference between subobjectives is revealed through the different rating partitionings for the Management Factor Items in that plot. Thus for subobjective 2, we can partition the space into three distinct areas containing points representing Management Factor Items which were rated 'High', 'Medium' or 'Low' in terms of their potential contribution to subobjective 2. The 11 different illustrations in Figure 3 (one for each of the 11 subobjectives) shows the mean categories of response for Management Factor Items in relation to each Subobjective. The plots are partitioned intoregions as shown to illustrate

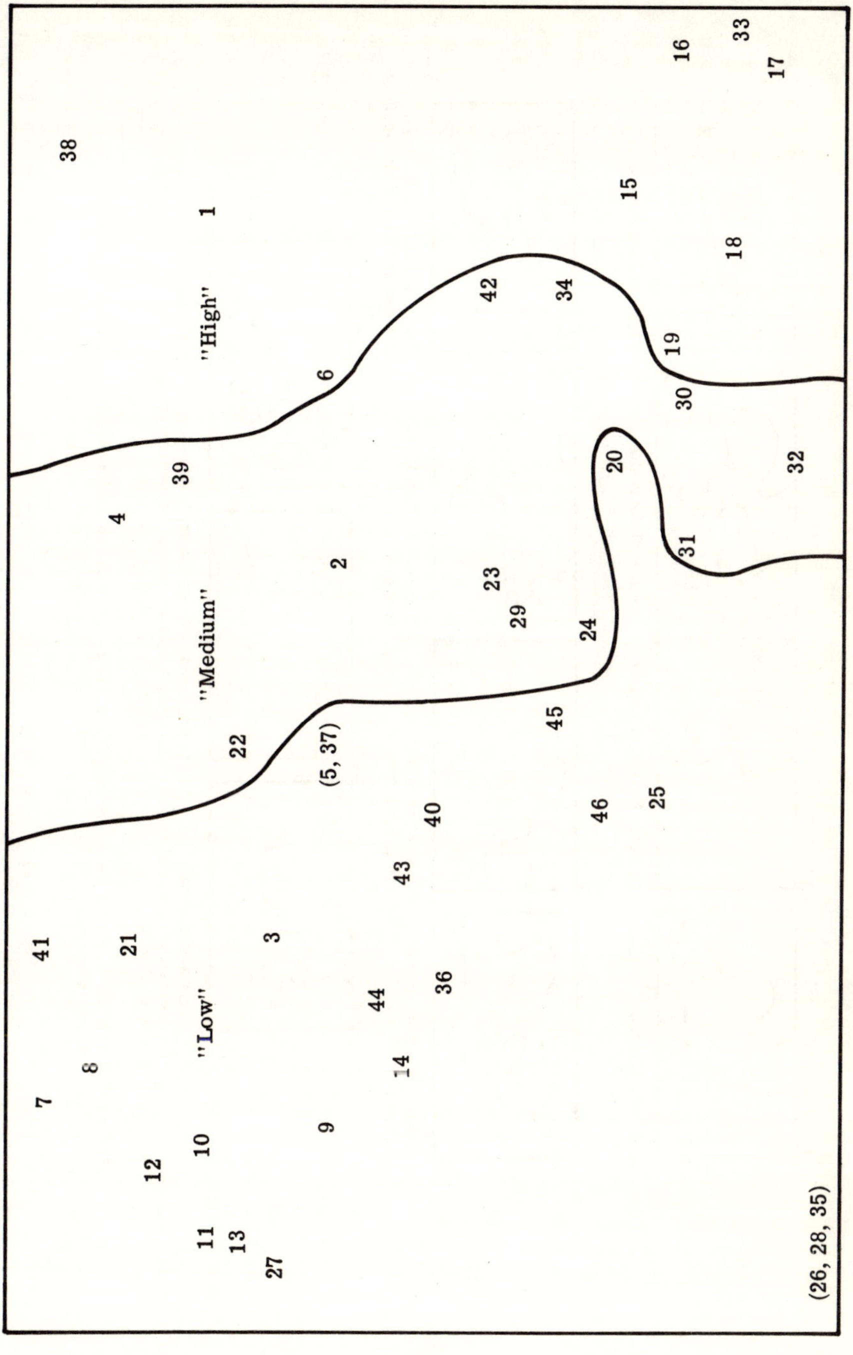


Figure 2: POSA Plot for Subobjective 2, 'Control of Corrosion and Erosion.

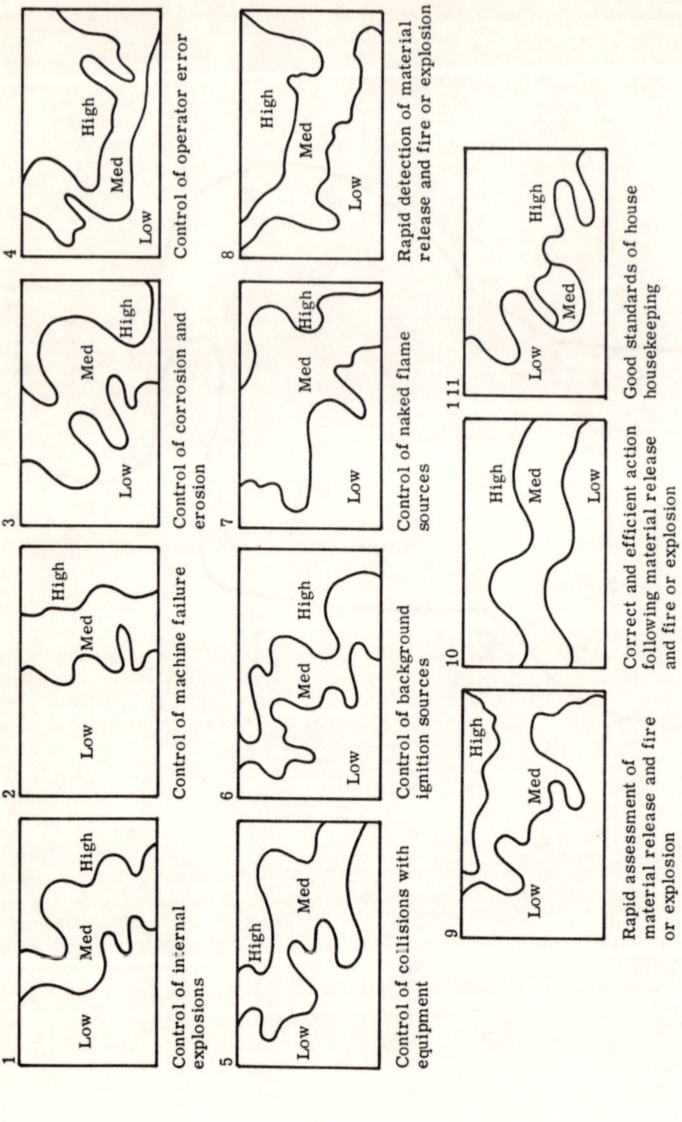


Figure 3: The Division of Management Factor Items into areas of Importance for the Eleven Subobjectives

the differences between the Subobjectives in terms of the way that the Management Factor Items are rated.

It is apparent that Subobjectives 1,2, and 3 have very similar patterns of response, whereas 4 and 5, although conforming to the same general pattern are less clear cut. A possible reason for this might be that 4 and 5 contain some reference or allusion to 'human action':

4. control of operator error
5. control of collisions with equipment

whereas 1,2 and 3:

1. control of internal explosions
2. control of machine failure
3. control of corrosion and erosion

are more concerned with the 'hardware or maintenance' of the process. Subobjectives 6 and 7 have similar patterns of response. Subobjectives 8,9 and 10 are rated similarly to each other and 10 exhibits a strong horizontal partitioning which can be strongly contrasted with the vertical partitioning of Subobjective 2.

The differences between the way in which Management Factor Items are rated in relation to Subobjectives, and the differences in terms of the 4 Primary Objectives are not as clear cut as may have been anticipated in terms of the Management Factor Structure. The strongest distinction to be drawn between Subobjectives is between those with an 'active/human' as opposed to a 'hardware/maintenance' component. The 'active/human' group can be further divided into those subobjectives concerned with general operator behaviour (4,5 and 7), and those concerned with human action in emergency situations (8,9 and 10). Control of standards of housekeeping really appears to be in a group of its own, and this might be expected in view of the fact that unlike other subobjectives, it does not deal with either human action or maintenance of the hardware/process, but is concerned with the working environment. This reasoning would give us 4 subobjective 'types' as hsown below:

1. maintenance of the process/hardware (1,2,3 and 6)
2. control of standards of housekeeping (11)
3. control of general operator behaviour (4,5 and 7)
4. control of human action in emergency situations (8,9 and 10)

With such a reduced number of Loss Prevention Subobjectives, the use of Primary Objectives would seem to be redundant. The Loss Prevention Subobjectives have the symbol 'C'. The overall outcome of this exercise was very useful in terms of providing a clear, more succint description of the Management Factor Items and the Loss Prevention Subobjectives, as well as indicating the potential value of the procedure for obtaining weightings for the Management Factor Items in relation to the Loss Prevention Subobjectives.

## FURTHER DEVELOPMENT OF THE MODEL

### Deriving the Loss Prevention Weighting Coefficients

At this stage two major components of the Management Factor Model, the Loss Prevention Subobjectives (C's) and the 12 groups of Management Factor Items which have been termed 'Management Factor Contributors' ($F_k$'s) had been identified and clarified. On the basis of the results of the previous exercise, we designed a modified questionnaire with the aim of producing a preliminary set of weighting coefficients for each Management Factor Contributor against each Subobjective. In order to do this, our representative Item was selected from each of the Management Factor Contributor Groups. The exercise took the form of 9 individuals being asked to rate the contribution made by each of these 12 Items to each of the 4 Subobjective on a 5-point verbal response scale. Participants in this exercise included experts from the chemical organisations we had visited as well as personnel from insurance bodies.

The results of this exercise were very encouraging since the judges did appear to use the entire response scale, no items were left uncompleted (indicating that the exercise was easily comprehensible and that the judges followed the instructions carefully), and there was a fairly high degree of consensus among the respondents. No unfavourable comments were received on either the content or the style of the questionnaire.

One way of analysing the data to produce the weighting coefficients would have been to simply find the mean of the rating scores given for each of the 48 subobjective/management factor contributor combinations. However, this would not have taken into account the variability in individual response patterns. The data was thus analysed using the Partial Order Scalogram Analysis technique as described in She and Elizur (8). Figure 4 shows the spatial layout of the 48 items (for all judges) as produced by POSA analysis. The items are coded from 1 to 4 for each of the Subobjectives and from 1 to 12 for each of the Management Factor Contributors. Thus item '3.11' relates to the contribution made by 'Human/Social Factors concerned with the State of the Organisation' to the 'Control of General Operating Behaviour'. The coding scheme is shown in Table 3.

In Figure 4, the distance of each of the 48 items along the diagonal axis provides a measure of its relative weighting: items 1.8 and 2.9 were rated as providing the greatest contributions, whereas items 1.12 and 3.1 were rated lowest. The remaining items are spaced fairly evenly along the axis, but with a slight tendency towards the centre.

TABLE 3 - Coding of the 48 Items in Rating Exercise

| Subobjectives | | Management Factor Contributors | |
|---|---|---|---|
| Control of: | | Executive Safety/State of Organisation | (1) |
| | | Executive Safety/Working Procedures | (2) |
| Housekeeping | (1) | General Safety/State of Organisation | (3) |
| | | General Safety/Working Procedures | (4) |
| Maintenance | (2) | Emergency Measures/State of Organisation | (5) |
| | | Emergency Measures/Working Procedures | (6) |
| General Operating Behaviour | (3) | Preventative Measures/State of Organisation | (7) |
| | | Preventative Measures/Working Procedures | (8) |
| | | Documentation/State of Organisation | (9) |
| Behaviour in Emergencies | (4) | Documentation/Working Procedures | (10) |
| | | Human/social factors/State of Organisation | (11) |
| | | Human/social factors/Working Procedures | (12) |

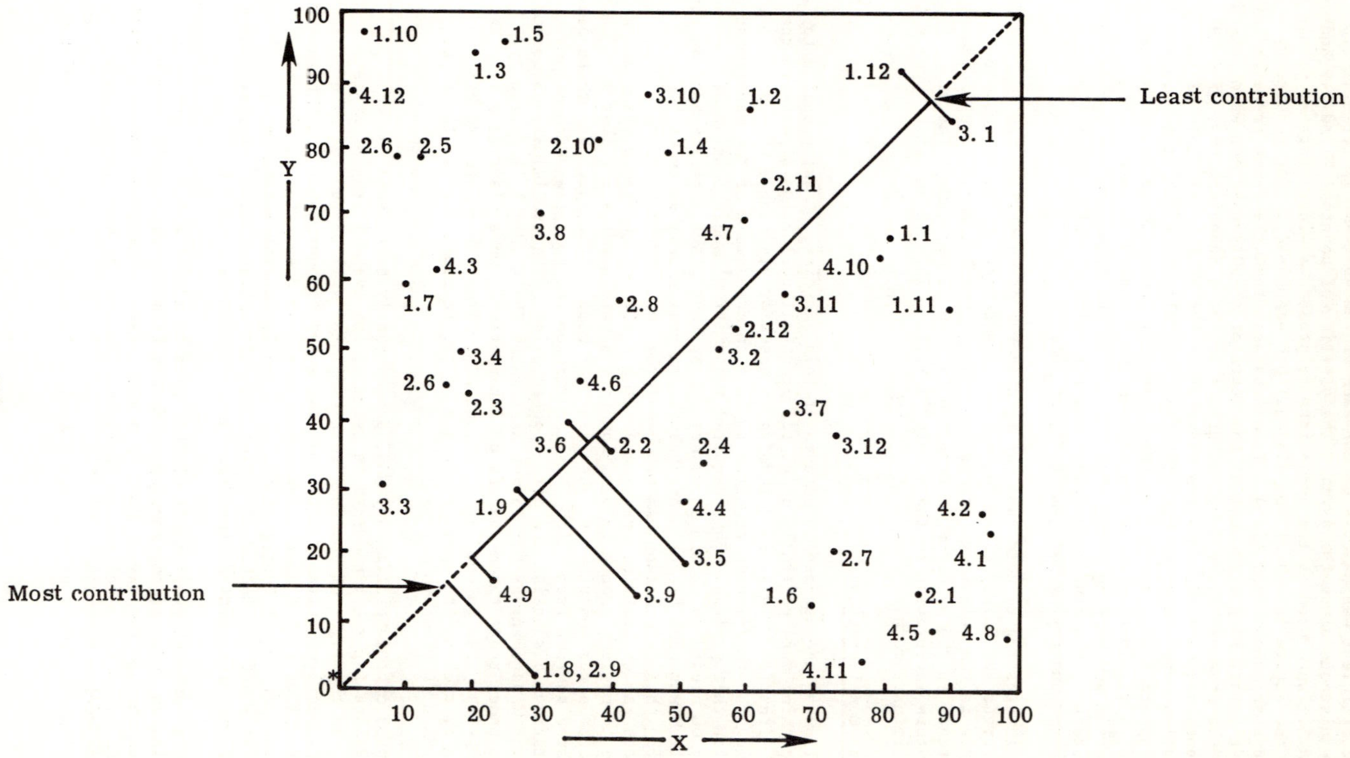


Figure 4: Spatial Layout of 48 Items in Expert Judge Rating Exercise as Produced by POSA Analysis.

13

A coefficient was calculated for each of the 48 items in terms of its relative distance along the axis. (The position of points 1.8 and 2.9 and of 1.12 and 3.1 were used as the end points to produce a wider range of values than if the entire length of the diagonal had been used). The resultant weightings are shown in Table 4. It can be seen that the items concerned with Executive Safety in the context of the State of the Organisation were judged to make the greatest overall contributions to the 4 Subobjectives, ($\bar{x}$=0.89). Items concerned with Preventative Measures in the context of State of the Organisation were judged to make the least overall contribution ($\bar{x}$=0.27). It is also apparent that apart from the Management Factor Contributor Emergency Measures, a distinction was drawn between the 2 contexts State of the Organisation and Working Procedures for all of the Management Factor Contributors. This finding supports the retention of this distinction in terms of the underlying Management Factor Model.

It is interesting to note that the mean scores of weightings for each of the Subobjectives are fairly close and towards the centre of the range. The difference in the weightings across the rows for each Subobjective indicates a difference in the patterns of response for the 12 Management Factor Contributors in relation to the 4 Subobjectives. This again lends support to the preliminary model.

## Practical Considerations

Following the successful outcome of this exercise, it was possible to begin to think more carefully about the practical collection of data in the applied situation. This would involve an assessor visiting a process site and collecting the information required to make an assessment of the Management Factor for that process. The 'required information' would consist of information needed to obtain an assessment of that plant's performance on each of the 12 Management Factor Contributors. A preliminary document for collecting this information has now been produced. For each of the 12 Management Factor Contributors, there are a number of questions which are designed to collect enough information to evaluate a value for that Management Factor Contributor. Use has been made of the fact that there are at least three sources from which this information can be collected: documentation, discussion and direct observation. The preliminary document, called the Management Factor Survey Assessment Chart (MASC) is currently arranged under these three sources of information. On a visit to a site, an assessor simply has to collect the required information and record his response by ticking the appropriate box on the response sheet. The first page of MASC is shown in Figure 5.

Copies of this preliminary MASC are currently being sent out to organisations for their comments in terms of its overall length, style and layout. In the light of comments received, MASC will be revised as necessary. Once MASC has been suitably refinded, the major components of the Management Factor Model will have been established; the Loss Prevention Subobjectives, the Management Factor Contributors, the Loss Prevention Weighting Coefficients and the method of collecting information (MASC). A preliminary model of the Management Factor Structure is shown in Figure 6.

A simple computer program has already been developed which is capable of evaluating an index of the Management Factor from the assessor's scores collected through the use of MASC on a site visit. There is, however, still much work to be done, before the Management Factor can be regarded as a viable measuring technique. Firstly, a function needs to be incorporated into the model to introduce process specificity. Initially some function of the

| Management factor contributors | Loss prevention subobjectives: Housekeeping | Process/Plant maintenance | General op. behaviour | Behaviour in emergencies | Means |
|---|---|---|---|---|---|
| Executive safety/state of organisation | .82 | 1.0 | .81 | .94 | .89 |
| Executive safety/working procedures | .51 | .39 | .29 | .21 | .35 |
| General safety/state of organisation | .83 | .58 | .45 | .32 | .55 |
| General safety/working procedures | 1.0 | .53 | .42 | .5 | .61 |
| Emergency measures/state of organisation | .2 | .25 | .36 | .65 | .37 |
| Emergency measures/working procedures | 0 | .44 | .43 | .59 | .37 |
| Preventative measures/state of organisation | .2 | .48 | 0 | .39 | .27 |
| Preventative measures/working procedures | .19 | .7 | .44 | .4 | .54 |
| Documentation/state of organisation | .43 | .78 | .97 | .69 | .72 |
| Documentation/working procedures | .33 | .63 | .77 | .67 | .6 |
| Human/social aspects/state of organisation | .4 | .6 | .73 | .55 | .57 |
| Human/social aspects/working procedures | .66 | .79 | .71 | .65 | .7 |
| Means | .46 | .60 | .53 | .55 | |

TABLE 4: Weightings Produced by the Expert Rating Exercise for Each Management Factor Contributor in Relation to each Loss Prevention Subobjective.

15

## I. Documentation

The following information may best be obtained through consulting the relevant records and documents. The most probable source for each answer is indicated in brackets following the question.

1. How often are the following individuals involved in safety committee meetings?

   (records of safety meetings)

| | always | usually | sometimes | rarely | never |
|---|---|---|---|---|---|
| senior managers | | ✓ | | | |
| middle managers | | | ✓ | | |
| line managers | | | ✓ | | |
| safety representatives | | ✓ | | | |

2. How well written is the company safety policy?

   (company safety policy)

| | |
|---|---|
| | very well written |
| | well written |
| | moderately well written |
| ✓ | not very well written |
| | not at all well written |

3. How comprehensive is the company safety policy?

   (company safety policy)

| | |
|---|---|
| | very comprehensive |
| ✓ | comprehensive |
| | moderately comprehensive |
| | not very comprehensive |
| | not at all comprehensive |

Figure 5: First Page of the Management Factor Assessment Survey Chart.

16

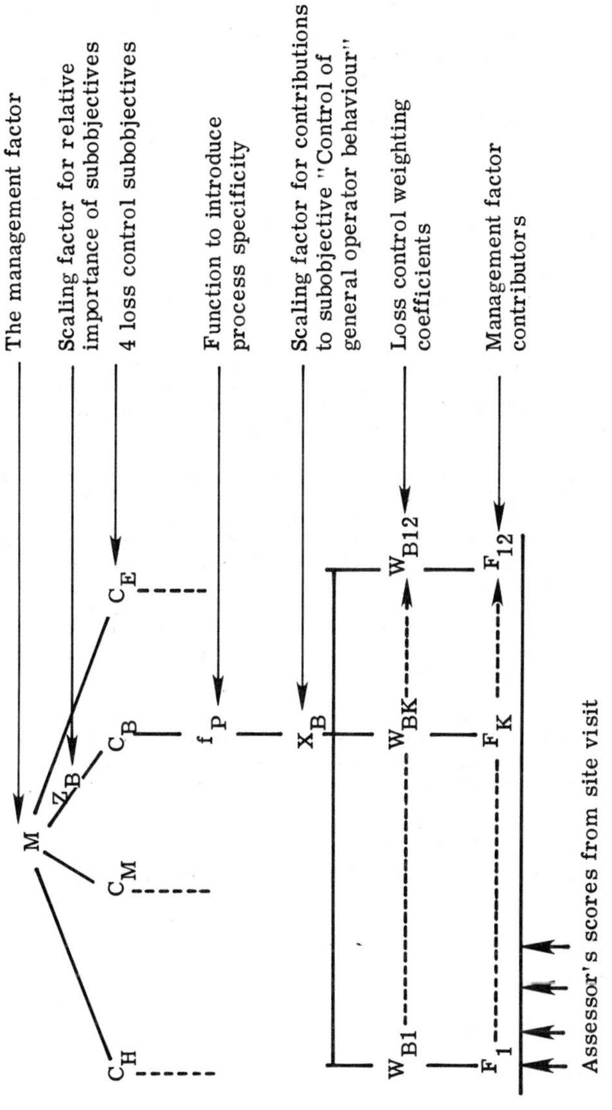


Figure 6: Preliminary Structure of the Management Factor Assessment Procedure.

17

IFAL 'p' factor will be used for this purpose. Secondly, we are not yet in a position to define the most appropriate range of values which the Management Factor can take. This will depend largely on the outcome of preliminary trials using MASC in the practical situation, and also on the form in which process specificity is finally introduced. At this stage, however, we are confident that the task of producing an effective tool for the assessment of the contribution made by the Management Factor to loss through fires and explosions in the chemical industry will soon be satisfactorily achieved.

## FUTURE POTENTIAL APPLICATIONS OF THE MANAGEMENT FACTOR TECHNIQUE

There are a number of other potentially valuable future applications of the Management Factor technique, some examples of which are given below:

1. As a research tool to investigate the effects of changes in external variables on a management's ability to control losses. It would be of great value to be able to assess the effects of, for example, changes in economic and unemployment markets on management standards and associated losses.

2. To investigate the effects of organisational size on the Management Factor and determine whether organisational size per se is associated with any particular patterns of weakness or strength in any of the management activity and states contributing to the overal Management Factor.

3. The Management Factor would be a very useful way of determining the effect of suggested practical improvements in management strategies on the overall ability of a management to control loss. A measure of the Management Factor could be made for a particular process, and in the light of this, suggestions could be made for the improvement of loss control activities in specific areas. Following the implementation of recommended changes in management strategy, the Management Factor for that process would be assessed again in order to determine the effect of the changes. Subsequent measurement of the Management Factor over time would indicate the permanence of any improvements.

4. The model underlying the Management Factor technique could easily be modified to assess standards of management in different industries (for example, the nuclear energy field) and to take into account other types of loss (personal injury or fatalities, business interruption costs, etc.).

## LIST OF SYMBOLS USED

CP = one of the 4 original Primary Loss Prevention Objectives (not used in current model)

$C_B$ = Subobjective "Control of General Operating Behaviour"

$C_E$ = Subobjective "Control of Emergency Behaviour"

$C_H$ = Subobjective "Control of Housekeeping"

$C_M$ = Subobjective "Control of Maintenance"

e = the engineering factor component of the IFAL algorithm

Fk = Management Factor Contributor (k=1 to 12)

f(p) = function of the IFAL process factor used to introduce process specificity into the calculation of the Management Factor

IFAL = Instantaneous Fractional Annual Loss

m(M) = the Management Factor

MASC = Management Factor Assessment Survey Chart

p = the process factor component of the IFAL algorithm

SGP = Standard Good Practice

wik = Loss Prevention Weighting Coefficient for the relative importance of Fk to Subobjective 'i', where i = H,M,B or E.

X = Scaling Factor

Z = Scaling Factor for relative importance of contributions to 'm' from $C_H$, $C_M$, $C_B$ and $C_E$

## REFERENCES


1. Parker, R.J., T (Chairman), 1975, "The Flixborough Disaster, Report of the Court of Enquiry", H.M.S.O. London.

2. Temple, C.J., 1977, Seveso, the Issues and the Lessons, Keith Shipton Developments Ltd., London.

3. Malone, T.B., Kirkpatrick, M., Mallory, K., Eike, D., Johnson, J.H., and Walker, R.W., 1980, "Human Factors Evaluation of Control Room Design and Operator Performance at Three Mile Island-2". U.S. Department of Commerce National Information Services, NUREG/CR, 1270, 3 volumes.

4. Cox, R.A. and Comer, P.J., 1980, "Development of Low Cost Risk Analysis for Process Plant", Institute of Chemical Engineers, Symposium Series no. 71.

5. Embrey, D.E. 1983, "The Quantification of Human Reliability using Expert Judgement: Current Findings and Future Developments", Paper presented at Symposium 'Human Reliability in the Process Control Centre' University of Manchester Institute of Science and Technology, April 20/21.

6. Singh, J., and Munday, G., 1979, "IFAL - A Model for the Evaluation of Chemical Process Losses", Design 79 Symposium, paper C-1, Institute of Chemical Engineering Midlands Branch September, 1979.

7. Canter, D., (1980/81), Fire Safety Journal 3. 41.

8. Romney, A.K., Shephard, R.N., and Nerlove, S.B. 1972. Multidimensional Scaling, Volume 2: Applications, Seminar Press, New York.

9 Shye, S., and Elizur, D., 1976, Human Relations 29, 63.

# NOTES

# ATTENTIONAL DEMANDS OF PROCESS CONTROL OPERATIONS: EFFECTS OF BOREDOM OR FATIGUE?

(R. Hockey, Department of Psychology, University of Durham)

One of the most serious consequences for human operators of the development of automated process control systems is the possibility of having to spend long periods of time in a state of general inactivity, only being called upon to act very infrequently. Such a working environment is monotonous and leads to boredom in most operators, though it is likely that the incidence of boredom will depend also on other characteristics, both of the task and of the operator. Boredom, in turn, is thought to be a potent source of inefficiency and loss of job orientation. What is less obvious, however, is that such situations may also give rise to problems that are related to fatigue.

Psychologists have often distinguished between boredom and fatigue in extended work situations. For example, Welford (1968) has attributed boredom to environments making too few demands on the operator for him to maintain an adequate state of alertness, while fatigue may be thought of as resulting from excessive work demands. This distinction seems acceptable intuitively, though there are reasons for doubting its value beyond the purely descriptive level. Like boredom, fatigue appears to be essentially a cognitive state, rather than, say, one characterised by physiological depletion. Further, again like boredom, it may be alleviated by a simple change of task requirements - not necessarily a reduction in task load - this applies even to fatigue resulting from continuous work of 24 hours or more! Holding (1983) argues convincingly that fatigue is best thought of as a strategic resistance to continued effort expenditure in attempting to meet current task goals.

The central point of this contribution to the workshop is that fatigue is a state which is not actually very different to that of boredom, but which captures more effectively the essence of the process monitoring task. It differs from boredom primarily, as Welford suggests, in terms of the degree of involvement of the operator with task information. However, both states may have their origins in the mismatch between preferred optimal deployment of mental resources and the information processing demands of the job. Studies of fatigue (such as those concerned with prolonged serial responding) have also revealed the presence of brief interruptions in the control of task-oriented behaviour, rather than gross reductions in baseline competence (e.g. Rabbitt, 1982). Furthermore, phasic effects of this kind, which can occur within the first few minutes of a work period, do not appear to be strategic. Rather they represent genuine drops in system performance - both accuracy and decision speed are adversely affected for a few seconds before normal levels of functioning are regained. Such an effect is referred to as one of fatigue largely because of the kind of task used to study it. The long time sampling required of traditional studies of monotony make it very unsuitable for investigating phasic lapses or blocks of this type. Yet, they may well be present in all monitoring situations (as well as others), in the form of rapid changes in attentional orienting (away from essential task elements) or sudden failures in information selection or intake.

Sustained monitoring of visual displays has traditionally been considered within experimental psychology in terms of watchkeeping or vigilance. These paradigms were based on original naval problems associated with watchkeeping, or related industrial processes such as inspection and quality control. They are based quite firmly on the monotony/boredom interpretation of the monitoring work situation, and may be less appropriate than has been supposed for the study of modern process control situations. Several broad problems of the watchkeeping model may be outlined:

(1) It assumes a relatively passive role for the operator, rather than the active involvement and intervention that may be required - parameter checking, control adjustments, information seeking, etc.

(2) It typically involves the detection of particular events or targets, rather than the broader goal of maintaining a system in an optimal state. This may involve attention to a large number of potential sources of information as well as internal monitoring of control procedures.

(3) Detection of faults is rarely the main problem in complex process monitoring tasks, since these are typically signalled by clearly audible alarms, as well as by visual indicators. More important is the diagnosis of the primary sources of failure in system components, and selection of the appropriate course of corrective action.

(4) Watchkeeping assumes little demand on auxiliary mental resources, such as those of memory and planning. Process monitoring requires anticipation about likely areas of difficulty and continual updating of current status. Advance information is often available, allowing the operator to monitor the development of the fault, before it is actually signalled.

It may be argued, with some justification, that such observations at best refer to the limits of applicability of the vigilance paradigm, rather than requiring something different. Indeed, current comprehensive reviews of the monitoring literature often include such data as exist on these issues as studies of vigilance (e.g. Davies and Parasuraman, 1979). This, however, only disguises the possibility that other methods may be more suitable as models of process monitoring. These should reflect, for example, the heavy decision making demands, problems of attentional control, the need to time share mental resources in the execution of elaborate search and checking strategies, and so on. The complexity of interactions of display information and control procedures forces the operator to develop a mental model of the status of the plant, and to use this model to actively plan his state, as well as to prepare possible appropriate corrective action in the event of a system failure.

We may still choose to think of the operator as being bored, and he may indeed confirm this by his subjective comments. Boredom is a subjective state, rather than an assessment of competence or efficiency. The methods for studying such effects in this situation, may, however, be best developed by considering them in terms of fatigue. Simply because the operator is not overtly busy all the time does not mean that he yearns for more task information to process - or that even if he did it would make him more efficient. In many complex installations the problems he will have to cope with are those of maintaining suitable selection, checking and control procedures in focal attention, rather than that of maintaining an adequate level of general alertness. We have to be very clear about the reasons for our concern over boredom. We should certainly try to make the operator's job more enjoyable and interesting, but we should not expect that this will also make him more reliable and efficient. Indeed in our efforts to achieve the first we may make it more difficult for him to achieve the second.

References

1. Davies, D. R. and Parasuraman, R. The Psychology of Vigilance. London: Academic Press, 1982.

2. Holding, D. H. Fatigue. In G. R. J. Hockey (Ed.) Stress and Fatigue in Human Performance. Chichester: Wiley, 1983.

3. Rabbitt, P. M. A. Sequential reactions. In D. H. Holding (Ed.) Human Skills. Chichester: Wiley, 1981.

4. Welford, A. T. Fundamentals of Skill. London: Methuen, 1968.

# THE OPERATOR IN A HIGHLY AUTOMATED PLANT: Some personal observations from experiences on Computer-controlled Batch Process plant, manufacturing fine organic chemicals.

(A. J. Faulkner, Organics Division,
Imperial Chemical Industries PLC, Huddersfield)

It is sometimes stated that boredom and monotony are problems which adversely affect the performance of Operators who supervise highly automated process plant. In considering this, it is necessary to define one's terms and consider whether the phrase "highly automated" can be properly applied to the situation. We assume that, when a plant is "highly automated" it actually runs itself, so that in such a situation the Operator is superfluous and a sensible management would, perhaps regretfully, dispense with his services. Then the problems of boredom and monotony do not arise in the sense that we are discussing it.

However, in real life, very few automation systems are actually capable of running a plant without human supervision. If the plant actually runs perfectly for 75% of the time, the humans are still needed to sort out the mess when the automation system goes beserk during the other 25%. Do they get bored when the plant is running well? The situation is at its worst for night shift workers in ways that are widely recognised, whilst day shifts are naturally busier because of all the day staff who have work to do on the plant. However, observations in the Fine Chemical manufacturing industry suggest that the problem is not as severe as might be supposed. There are three main reasons for this:-

(i) There are always plant operations which cannot be automated reliably or cost effectively. In this case, the automation system will relieve the Operator of the monotonous routine parts of the operation of the process, but, at times, it has to suspend itself whilst the Human carries out the difficult manipulation. Typical areas in which this occurs in Batch manufacture of Fine Chemicals are:-

(a) The charging or discharging of solids from, or into, non-bulk packages.

(b) Taking samples and carrying out complex qualitative and quantitative tests for Quality Control purposes, at the required intervals throughout the batch cycle.

(c) Situations in which instrumentation cannot be expected to work reliably, e.g. interface detection in liquid phase separations.

(ii) The nature of the chemicals involved (e.g. corrosive, abrasive slurries) results in plant reliability problems, with the consequent need for Human intervention to protect the batches against decomposition yield losses, etc., and yet make the plant safely available for the maintenance workers to repair it.

(iii) The chemical processes involve the synthesis of complex organic molecules, and because of the continual search for improved products and greater efficiency, development is always going on. This means that there are often new situations and new operating problems which the Operators have to deal with.

Of course, it is the task of the Process Engineers and Development Chemists to reduce the adverse effects of these three factors on the efficiency of the plant operation, but bearing in mind the ever-changing nature of the business and the difficulty of some of the problems, a situation in which our type of plants and processes are fully automated and 100% reliable is still a long way off.

Another fact of life in this decade caused by the general search for cost savings has been the overall reduction in manning levels throughout the Industry. This has prompted Managements in search for efficiency and cost saving, to re-structure jobs in order to make better use of the skills available. As a result, for example, Shift Managers look after larger areas and Senior Operators take on some management responsibilities, becoming Group Leaders whilst still carrying out their duties as Operators. As a consequence, they don't have time to be bored for long.

In some industries, particularly where green-field sites are concerned, agreements have been achieved which permit the relaxation of traditional trade demarcations, and these give even greater scope for enhancement of the Operator's job, as he can be trained to take on, for example, some of the surplus plant maintenance tasks, which are compatible with his routine work.

In these ways, and no doubt in many others, external factors have led to organisational decisions which as a sort of common-sense by-product enhance the Operator's job and significantly reduce the possibility of boredom.

Perhaps there is a lesson here for industries which appear to have a problem with Operator boredom. A bored operator is an intelligence going to waste. The problem disappears if ways can be evolved to utilise effectively this valuable resource. The effort in re-education, training and in modifying job structures will be well worthwhile.

NOTES

NOTES

NOTES

Printed by The Chameleon Press Limited, 5–25 Burr Road, London SW18 4SG